Information Concepts

From Books to Cyberspace Identities

Information Concepts: From Books to Cyberspace Identities

Gary Marchionini

`www.morganclaypool.com`

ISBN: 9781598299625 paperback
ISBN: 9781598299632 ebook

DOI 10.2200/S00306ED1V01Y201010ICR016

A Publication in the Morgan & Claypool Publishers series
SYNTHESIS LECTURES ON INFORMATION CONCEPTS, RETRIEVAL, AND SERVICES

Lecture #16
Series Editor: Gary Marchionini, *University of North Carolina at Chapel Hill*
Series ISSN
Synthesis Lectures on Information Concepts, Retrieval, and Services
Print 1947-945X Electronic 1947-9468

Synthesis Lectures on Information Concepts, Retrieval, and Services

Editor
Gary Marchionini, ***University of North Carolina at Chapel Hill***

Synthesis Lectures on Information Concepts, Retrieval, and Services is edited by Gary Marchionini of the University of North Carolina. The series will publish 50- to 100-page publications on topics pertaining to information science and applications of technology to information discovery, production, distribution, and management. The scope will largely follow the purview of premier information and computer science conferences, such as ASIST, ACM SIGIR, ACM/IEEE JCDL, and ACM CIKM. Potential topics include, but not are limited to: data models, indexing theory and algorithms, classification, information architecture, information economics, privacy and identity, scholarly communication, bibliometrics and webometrics, personal information management, human information behavior, digital libraries, archives and preservation, cultural informatics, information retrieval evaluation, data fusion, relevance feedback, recommendation systems, question answering, natural language processing for retrieval, text summarization, multimedia retrieval, multilingual retrieval, and exploratory search.

Information Concepts: From Books to Cyberspace Identities
Gary Marchionini
2010

Estimating the Query Difficulty for Information Retrieval
David Carmel, Elad Yom-Tov
2010

iRODS Primer: Integrated Rule-Oriented Data System
Arcot Rajasekar, Reagan Moore, Chien-Yi Hou, Christopher A. Lee, Richard Marciano, Antoine de Torcy, Michael Wan, Wayne Schroeder, Sheau-Yen Chen, Lucas Gilbert, Paul Tooby, Bing Zhu
2010

Collaborative Web Search: Who, What, Where, When, and Why
Meredith Ringel Morris, Jaime Teevan
2009

Multimedia Information Retrieval
Stefan Rüger
2009

Online Multiplayer Games
William Sims Bainbridge
2009

Information Architecture: The Design and Integration of Information Spaces
Wei Ding, Xia Lin
2009

Reading and Writing the Electronic Book
Catherine C. Marshall
2009

Hypermedia Genes: An Evolutionary Perspective on Concepts, Models, and Architectures
Nuno M. Guimarães, Luís M. Carrico
2009

Understanding User-Web Interactions via Web Analytics
Bernard J. (Jim) Jansen
2009

XML Retrieval
Mounia Lalmas
2009

Faceted Search
Daniel Tunkelang
2009

Introduction to Webometrics: Quantitative Web Research for the Social Sciences
Michael Thelwall
2009

Exploratory Search: Beyond the Query-Response Paradigm
Ryen W. White, Resa A. Roth
2009

New Concepts in Digital Reference
R. David Lankes
2009

Automated Metadata in Multimedia Information Systems: Creation, Refinement, Use in Surrogates, and Evaluation
Michael G. Christel
2009

Information Concepts

From Books to Cyberspace Identities

Gary Marchionini
University of North Carolina at Chapel Hill

SYNTHESIS LECTURES ON INFORMATION CONCEPTS, RETRIEVAL, AND SERVICES #16

ABSTRACT

Information is essential to all human activity, and information in electronic form both amplifies and augments human information interactions. This lecture surveys some of the different classical meanings of information, focuses on the ways that electronic technologies are affecting how we think about these senses of information, and introduces an emerging sense of information that has implications for how we work, play, and interact with others. The evolutions of computers and electronic networks and people's uses and adaptations of these tools manifesting a dynamic space called cyberspace. Our traces of activity in cyberspace give rise to a new sense of information as instantaneous identity states that I term proflection of self. Proflections of self influence how others act toward us. Four classical senses of information are described as context for this new form of information. The four senses selected for inclusion here are the following: thought and memory, communication process, artifact, and energy. Human mental activity and state (thought and memory) have neurological, cognitive, and affective facets. The act of informing (communication process) is considered from the perspective of human intentionality and technical developments that have dramatically amplified human communication capabilities. Information artifacts comprise a common sense of information that gives rise to a variety of information industries. Energy is the most general sense of information and is considered from the point of view of physical, mental, and social state change. This sense includes information theory as a measurable reduction in uncertainty. This lecture emphasizes how electronic representations have blurred media boundaries and added computational behaviors that yield new forms of information interaction, which, in turn, are stored, aggregated, and mined to create profiles that represent our cyber identities.

KEYWORDS

information, cyberidentity, human-information interaction, proflection

Contents

Preface

Information is a term that conjures different meanings depending on who uses it and in what context. This lecture provides an overview of several senses of information in use today and introduces a new sense that has arisen as a result of the dynamic interactions of large numbers of humans using global communication networks that are embedded with a variety of algorithms that facilitate, monitor, and analyze these interactions. This lecture grows out of empirical studies of information seeking conducted by the author and others over the past thirty years and observations of the evolution and current state of what has come to be known as the World Wide Web (WWW), which together with the human and computational entities that use it is called cyberspace.

The intention here is not to discuss every sense of information in depth, but rather to provide general introductions to some of the ways that scholars have treated the concept in order to provide a base for understanding how we interact with ideas, people, and electronic artifacts in cyberspace and what informational effects these interactions create. For each of us, these interactions are myriad and persistent projections of our selves. People, institutions, and computational agents leverage these projections for many purposes, which are themselves myriad and persistent and serve to reflect our identities. The evolving projections and reflections represent our cyber identities, which I call proflections of self. Our proflections are meaningful because they impact how people act toward us and how we act in turn.

Gary Marchionini
October 2010

Acknowledgments

This lecture was influenced by many colleagues, especially my students. Over the last decade, I have offered a challenge to incoming doctoral students to become invisible in cyberspace and then write how they do this as a dissertation. Fred Stutzman, Terrell Russell, Chirag Shah, Ben Brunk, Jung-Sun Oh, Javier Velasco-Martin, Sanghee Oh, and Jacob Kramer-Duffield each considered the challenge and took interesting tacts on issues of social media and taught me much about these environments. Barbara Wildemuth, Robert Fu, Yaxiao Song, Meng Yang, and Gary Geisler helped me to grapple with issues of information seeking in WWW-based multimedia, and Cassidy Sugimoto led a team that developed a database (MPACT) of social network relationships among doctoral students and their mentors (advisors and committee members). My colleagues at UNC have always been excellent sounding boards, and I thank them for their patience with me as I fumble with new terms to describe the phenomena we study and teach.

I would also like to thank several colleagues who read drafts and made many helpful suggestions, especially Ben Shneiderman, John Tait, Michael Buckland, Ian Ruthven, Fred Stutzman, and Rob Capra. Although this lecture benefitted from their comments, I did not take all the advice they gave and all the errors and omissions are due to me alone.

Finally, my wife Suzanne is due a debt of gratitude for her support and the missed attention I shunted to this lecture.

Gary Marchionini
October 2010

CHAPTER 1

The Many Meanings of Information

Information pervades our lives. Whether executing simple acts or making complex decisions, we depend on information to perform effectively. For people, information serves to sustain our mental activity just as food and drink sustain our bodies. Likewise, economic and social organizations depend on information to function effectively. The information used and created by people and institutions serve as expressions of commerce and culture. More concretely, physical, chemical, and biological processes use information to effect change in physical states. Clearly, something that applies to so many fundamental activities will have a variety of manifestations and interpretations. This lecture considers several of the most important senses of information and introduces a new sense made possible by ubiquitous, massively social computational systems and the collective behavior of people within these systems. These systems and human activities with them comprise what is called cyberspace.

The pioneering work of Fritz Machlup applied economic measures to information production and use and introduced the concept of the information society to the post-WW II era (e.g., Machlup, F. [1962]). By 2006 in the U.S. alone, more than 3.3 million people worked in a trillion dollar information industry according to the U.S. Statistical Abstracts[1]. More and more of the new information products are born digital, and a recent New York Times article reported 281 exabytes of data online in 2009 with more content expected to be uploaded in the next four years than all the content created in human history[2]. These factoids drive home the importance of information in our world and motivate consideration of the implications of these trends for our personal and collective lives, and more broadly for our institutions, cultures, and species. Understanding the phenomenon called 'information' is crucial to human progress and mutual understanding in a highly connected and resource-limited world.

In this chapter, I introduce five different senses of information via countervailing extreme points of view and then offer a framework for classes of interpretation. The chapter closes with a discussion of information from a human perspective that serves to scope the overall framework and chapters that follow.

[1]The 2010 Abstracts are based on 2006 data: `http://www.census.gov/prod/2009pubs/10statab/infocomm.pdf`.
[2]Macmanus, R. May 31, 2010. New York Times.

1.1 FIVE SENSES OF INFORMATION

What do we mean when we use the word 'information?' As is common in human language, most words have several meanings and uses. To set the stage, I begin with the rhetorical device of a forum that gives platform to five particular points of view that represent four classical senses of information and introduces one new sense that arises from the traces of people working in cyberspace. These five senses are treated in respective chapters with emphasis on how electronic digital technology has affected each the four classical senses and given rise to cyberspace identity as a new form of information.

These five different fictitious voices are meant to illustrate very different ways to think about information. They also set the stage for discussion of new senses of information that arise as electronic technologies are more integrally coupled to human minds and bodies, to the artifacts humans create, and to the physical and electronic environment that surrounds us. They do not represent all senses of information but serve to contextualize the emergence of the new senses that arise in cyberspace.

Voice1: Thought and Memory. Information is a state of the human mind. *Information exists only in the mind as an activated state of consciousness and has no place outside the mental world. This state occurs as recalled memories are instantiated in an instantaneous life context. Often, the memories are recalled in response to some external set of stimuli from the environment—we make meaning from the stimuli. For example, at the theater, a particular utterance or action by an actor may trigger a set of memories and impressions. The overall mental state is concurrently affected by the events of the day and one's current understanding of the world. One's memory traces and interpretations of the world are thus information that define and compose one's state of mind at any instant of time. In some cases, we establish an information state of mind without external signals by reflective thought. Thus, information is the 'stuff' of thought; the objects in our working memories that allows us to think and guide behavior.*

This comprises a cognitive view of information in that things external to the human mind are not information but signals, data, physical states, or information artifacts. Books or websites or video games are not information, rather, the set of memories and thoughts that they stimulate are information. One may perceive signals from the world but choose not to interpret them (in fact, selective filtering is essential to avoid overloads that cripple thought) and unless one attends to the signal, thus changing mental state, the signals do not stimulate information.

Voice2: Communication Process. Information is a stimulus perceived by a human; it is the process of affecting mental state. *Voice1 confuses thought with information and smacks of solipsism. Information is a signal from outside the brain that is perceived by a range of receptors and processed by a brain. In human communication, it is a message sent and received with intention. Interpretation and thought are distinct from information, although they typically use information. The information at the theater lies in the light and pressure waves reaching one's eyes and ears. The resultant neural activities allow one's brain to make sense of the actor's words and actions. It is the active stimuli and the human perception of the stimuli that define information. The stimuli not perceived by a human are signals but not information. A tree that falls in the forest causes a signal, but it is not information unless some human perceives the signal.*

This comprises a strongly physiological process view of information; things not perceived by humans are merely physical signals or artifacts that have the potential to become information upon human perception. Likewise, the mental activity that results from information perception is not itself information but thought or cognition that uses information. Some who adopt this point of view focus exclusively on the stimuli that come from other people as information (what some term communication) and others consider all stimuli from the natural world as information.

Voice3: Artifact. Information is matter that has been manipulated by people or their computational agents—artifacts that people create to communicate. *The other voices are all interesting academic arguments that do not really have much impact on life. Voice1 confuses information with thinking, Voice2 confuses it with perception and communication, and Voice4 (below) confuses it with physics and communication theory. Information is not the meaning one attaches to the actor's words or actions or the process by which they get to our heads, but rather the script, choreography, and any record of the performance. Although this information may be ephemeral as in the stage performance, more often it is proscribed in a physical artifact such as paper, magnetic charges, or pits on stone or metal. Information is the book or newspaper, webpage or sensor stream as a real object, not the interpretation in a human mind when reading or the act of delivering the object. Information is tangible and measurable. It is produced by people for a purpose and can be copied or destroyed.*

This voice comprises a highly practical view of information as matter that is created and managed. This sense of information gives rise to database, media, and publishing industries as well as cultural memory and educational institutions like libraries, museums, and archives.

Voice4: Energy. Information is a kind of energy. *Information is a fundamental force of nature. The debate is whether it is the potential capacity to change mental or physical states, the actual force that effects transitions between these states, or the effects of these state changes. Because we cannot measure the energy itself, we measure the work it does—the state changes—as measures of information. There is nothing inherently human about information, although humans are the most complex information users on the planet and thus often appropriate information to their own activities. In the theater example, the actor's posture, position, movement, and the pressure waves emanating from the vocal chords change the physical state of the theater, and these changes, in turn, propagate to the audience who use their perceptual system to gather this energy and change their individual mental states. Thus, the theater is full of information.*

This view is more basic than Voice1 because the state that may be changed could be a mental state in a human but also could be the state of a cell in a frog or the orbit of a communication satellite. Mental activity is the epitome of information as energy, but there are many forms of information as energy at less complex levels. Objects considered by Voice3 such as books, disk drives, and websites are all examples of potential energy—able to be activated by secondary processes to change mental or other states. Likewise, this view is more basic than Voice2 because the perceptual signals are kinds of kinetic energy that change the states of the human peripheral and central nervous system. Moreover, it is broader because it allows other organisms, machines, or other phenomena to use information.

Voice5: Identity in Cyberspace. Information is an instantaneous state of cyberspace. *Voice1's view of information in the head has been augmented by external manifestations of thought enabled by rapidly accessible 'cloud-based' storage and inferential tools ranging from data analytics to visualization. Voice2's concern with information acts has been exponentially magnified by global networks that amplify personal and organizational voices both spatially and temporally. Voice3's view of artifacts and industries that have arisen to produce, distribute, and manage them over the ages has been disrupted by new technologies for creation, distribution, and management. Voice4's abstract conception of energy is exacerbated by huge increases in potential and kinetic values that global networks enable. All these traditional senses of information have been disrupted by electronic technologies. More importantly, the respective disruptions give rise to entirely new forms of information that have theoretical and practical import. Most notably, the state of cyberspace at any instant is a new kind of information and particularly important partitions within this cyberspace exist at any instant for every person who works or plays online. These partitions are important because they represent our personal identities in cyberspace and are called proflections of self.*

Each of these admittedly bombastic and strained points of view has some element of resonance with how people think about information. They do not exhaust the possible views and have many variants and in practice tend to be combined. Here they are meant to serve as advance organizers for what follows. Voices1, 2, and 3 are strongly human-centric, although where the information lies in the human ranges from strictly cognitive (Voice1) to strictly physical (Voice2) to strictly external (Voice3). Voice4 is broadest and aims to treat information as a ubiquitous phenomenon. Voice5 is the motivation for this lecture. Voice1 is concerned with information in the mind and Voice2 with information getting to the brain. Voice3 and Voice4 are concerned with information in the environment, although Voice3 focuses on the products created by people. Voice1 is strongly affiliated with cognitive science, Voice2 with physiology and communication behavior, Voice3 with engineering and data management, Voice4 with mathematics and physics, and Voice5 with new meanings for information in the 21st century and the roles that information plays in our increasingly connected lives.

Information scholars have proposed general definitions of information. Schrader, A. [1984] surveyed two score different senses of information studies[3]. Losee, R. [2010, 1997], argues that information is the product of a process and uses this approach to subsume discipline-specific senses of information. This argument is much like Voice4 in that the product of a process (state change) is defined as the information. Others (e.g., Buckland, M. [1991b]) have considered three main senses for the term 'information.' One goal of this lecture is to argue that electronic media are blurring these distinctions, which, in turn, is causing new meanings and roles for information to emerge.

Resnikoff, H. [1989] argues that the science of information is rooted in four disciplines that led to the development of computers and the information processing model of psychology: thermodynamics (entropy), psychophysics (selective attention), quantum mechanical theory of measurement (quantization), and communication engineering (bandwidth and coding).

[3]The focus here is on information as a phenomenon rather than the nature of information science, informatics, information studies, or library science as schools or fields.

A more human-centered philosophy focuses on semantics that reflect personal and social meaning making. Karamuftuoglu, M. [2009] makes the case for an epistemological perspective for information that focuses on how and why people create, interpret, and use information. This view is also reflected by Hjorland, B. [1995] who argues for a social rather than cognitive view of information.

A historical perspective on information is represented by Hobart and Schiffman [1998] who trace the interaction of technology and culture over three ages of information from literacy that emerged 5000 years ago (written language was the first language technology) to numeracy (mathematical technology) that drove analysis and empirical science, to the computer age that adds computational technology to give new meanings to information as the product of technology and culture. Furner, J. [2004] argues that all the senses above already have words and concepts associated (e.g., message, communication, relevance) and there is no unique meaning for the term information. These are just a few of the ways that scholars have chosen to elucidate what we mean by information and illustrate the broad scope and fundamental nature of information in our world today.

In this lecture, I will use a socio-technical perspective to argue that electronic media and cyberspace, which are fast becoming the dominant means for representing information outside the head, cause very tight coupling of the classical senses of information. This tight coupling, in turn, gives rise to extensions of our selves that propagate far beyond our intentions or control. When we interact with electronic information, we become part of it and leak some of our selves (exoinformation) into the electronic information space. The emerging and continually morphing electronic information space is termed 'cyberspace,' and it is enabled by global electronic networks that are instantaneous, incorporate computational agents (programs), and they are becoming massively social. Cyberspace will be characterized in more detail in later chapters, but here it is useful to note that the cyberspace includes the actions of people and the subsequent traces of those actions enabled by its technical components. Moreover, although I will contrast information in mental space, physical space, and cyberspace, this does not imply that cyberspace experience is any less real than what takes place in our minds or in the physical world.

The intentional or unintentional extensions of our selves into cyberspace along with social and computational actions related to these extensions is termed *proflection* and results from conscious and unconscious *pro*jections plus human and computational agent re*flections*. Thus, proflection of self in cyberspace is a new kind of information and will be included in the framework that follows as a fifth sense of information.

1.2 INFORMATION SENSES FRAMEWORK

Figure 1.1 depicts five different kinds of senses for information that are treated in more detail in this lecture. Thought and memory and communication processes are treated as strictly human-centric senses and map to Voice1 and Voice2, respectively. Information as artifact traditionally has meant manifestations of thought or communication in physical objects. In our treatment here, we expand the manifestations to electronic forms that require other kinds of artifacts (hardware and software) to allow persistence (storage) and human perception–an extension of Voice3. Energy is the most general sense of information and is expressed mainly in Voice4. Just as there are many different types of energy treated in physical sciences, each with distinct properties and measures, there are different kinds of informational energies, most notably, analogs to potential and kinetic energy and the resulting work they manifest. In human-centric information science, the work done is mental or physical state changes. Physical state change measures of information provide the basis for classical information theory, and mental state changes provide the basis for human-centric senses of information. The final sense, cyberspace state and more specifically, proflection of self in cyberspace (Voice5), is a recent phenomenon rooted in the classical senses of information and amplified by computational social networks.

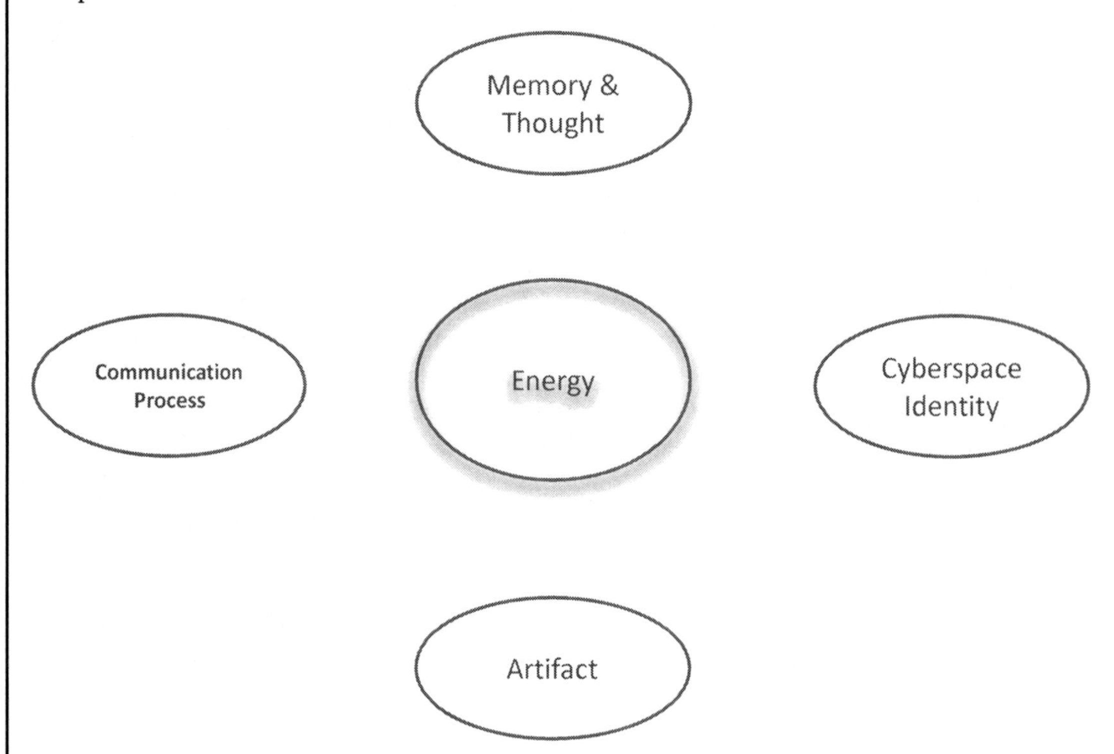

Figure 1.1: Information Senses.

In the figure, energy is central because it is the most general sense of information and can be related to the other senses. External (to the human body) energy and internal memory activations drive thought as a stream of mental state changes (Voice1). Physical energies external to the body stimulate biological state changes in the brain (Voice2). Potential energy supports information as artifact (Voice3). Cyberspace represents the sum of human and machine communication processes and artifacts. It is a subset of all human knowledge augmented by products of computational agents, which, in turn, has implications for individual identity and for collective consciousness.

1.3 INFORMATION TERMINOLOGY AND THE SOCIO-TECHNICAL PERSPECTIVE

This lecture is about information with a focus on a new sense of the concept, but there are many related terms that should be distinguished. It is common to see the terms 'data,' 'information,' and 'knowledge' used interchangeably. Clearly, these are related concepts; however, it is useful to be more precise and distinguish these terms. Taking information as a kind of energy, we can characterize noise, data, information and knowledge, increasingly ordered and complex energy forms, with new qualities emerging as order and complexity increase. Generally, the new qualities support larger or broader state changes when the energy is activated. Order can result from physical constraints that channel the flow of energy, from systematic varying of intensities, or from sequenced interruptions of energy flow. Complexity of energy is due to the number of variations in the ordering principles applied over time or from combining multiple energy flows.

Random energy is the absence of information. Random energy manifests without predictable pattern (e.g., Brownian motion). There is an important complementary absence of information defined by continuous and unchanging energy—an always present and unvarying energy state. This state serves as the signal: the ground, carrier, or medium for other energy flows.

Energy that has some order is known as data. Data manifest in many forms determined by the ordering constraints. Consider a heat sensor that sends a stream of readings to a remote monitoring station. The stream of readings illustrates one kind of data. One specific reading at a particular point in time is a datum. Within the sensor, the physical reading is a specific point in a range of sensitivities built into the sensor. Say that the sensor can detect 1000 different heat conditions, the heat in the environment at a particular time interval changes the sensor state to one of the 1000 positions on the sensor—say 632. This datum represents the heat energy in the sensor's environment and must immediately, in turn, change the state of a display device, a storage unit, or a transmission channel, else the unit of time reading passes and the temperature for the next time interval is represented in the sensor. We can see there are fine transductions between the instantaneous heat energy of interest and the record of this energy. The record, in turn, has the potential to change various system states. If the sensor is to be used locally, then it may take the 632 datum and map it to some standardized scale, for example, degrees Celsius and change the display state to reflect that datum, for example, 99.5. The 632 datum may be represented in various storage forms (e.g., as 1001111000 in a flash memory unit) or the value may be transduced for transmission purposes as a stream of radio waves. A reading at any given instant is a datum that is systematically mapped to a code, and stored, or it disappears. Thus, data that are displayed, stored, or transmitted have the potential to change the states of atoms or minds. Data have no meaning to humans without minimal metadata such as units of measure.

Energy that has order as well as associated energy that provides context is called information. The contexualizing information is metadata that supports the flow and use of data. For example, sodium ions in the human body change cellular states by passing through channels that upregulate or inhibit according to electro-chemical conditions in a cell membrane. The same ion has no effect out of the context of the associated biological pathway. A flag flown at half mast changes an observer's mental state by virtue of its fixed but abnormal position on the pole. In the case of changing human minds, information is interpretable as a result of the metadata and the associated mental activity that integrates these data.

Energy that originates in social contexts and that organizes different information resources into aggregates according to socially accepted schemas is called knowledge[4]. The schemas can be temporal, spatial, or conceptual and serve to connect information nodes in systematic ways. Knowledge entails both declarative (concept-based) and procedural (how to) components and in the human individual or organization includes both tacit and implicit capabilities. Knowledge tends to have substantial scale and includes context for the components and relationships. Knowledge has explanatory and predictive power and tends to be associated with humans or human organizations. When people develop techniques or systems to apply knowledge, the field of knowledge management obtains. See Alavi and Leidner [2001] and Koenig and McInerney [in press] (this series) for introductions to knowledge management.

The term 'wisdom' is sometimes included in this data-information-knowledge hierarchy; however, I reserve it for something distinctly human and singular that is of a different phenomenon class than the data-information-knowledge hierarchy of increasingly ordered, contextualized, and expansive energy. Wisdom is an inherently human characteristic that entails intuition, trust, empathy, patience, humility, and personal and social balance. Whether wisdom is due to experience or to chemical balances in the body, it is not dependent on information at hand or at mind but perhaps by what we are able to do with and without information. Hall, S. [2010] examines the meanings that scholars have given to wisdom over the years. Early philosophy related wisdom to experience rather than intelligence. Twentieth century work began to associate wisdom with aging and the ability to recognize life-important rather than intellectually important situations and to balance intelligence and emotion.

One other terminological caveat will be helpful in this lecture. I use the term electronic digital artifact rather than the more popular digital artifact. Digital artifact in popular usage assumes that one or more electronic computational devices are involved, however, to be precise, we can create digital artifacts on paper or other substrates, and we have many electronic analog artifacts, so I use more precise term 'electronic digital information artifact.'

This chapter serves as an organizing framework for the following chapters that consider the different senses of information, in turn, with an eye toward how these senses are influenced by electronic systems and with particular emphasis on a new kind of information that has arisen as a result of humans using electronic systems for a half century. We have entered a critical transition stage of the evolution of information and the human experience. Billions of people do what is termed information work and social media connect physically disparate individuals and communities. We have transitioned from the early adoption stage to the point where electronic information is the norm rather than exception. This transition causes us to reflect on the implications of our adoptions

[4]The meanings and literature of 'knowledge' is larger and more diverse than 'information.'

and the adaptations that inevitably follow. Tensions between efficiency and privacy, ownership and reuse, selection and exhaustion, expertise and crowd wisdom, creativity and homogenous standards, and personal and public essences abound. It is an exciting and defining time for information theorists and professionals.

CHAPTER 2

Information as Thought and Memory

Information as thought and memory considers where human-generated information originates and where it typically finds usage. At a micro level, information in the head is the energy in collections of synapses that are concurrently activated. At a more practical scale, as Voice1 (mental) argues, information in the head is the set of concepts and relationships active at a given time interval. This condition may arise through introspection and reflection on concepts or events recalled from memory, or it may arise through external stimuli acquired by our perceptual system. The most common sense of information in the head is the mental state that results from an interpretation of an external stimulus, whether from the ambient environment or an information artifact upon which we have focused our perceptual system. An extreme variation on Voice1 claims that everything external to the human mind is data or signals but information only exists within the human mind. The various notions of information as thought and memory are considered here in three general classes, information as what we know (cognition that tends to be associated with prefrontal brain activity), information as how we know, and information as what we feel (emotions and intuitions that tend to be associated with the amygdala).

2.1 NOUMENAL CLOUDS

From a conceptual and somewhat metaphorical perspective, consider that new information originates in human minds as a function of what we commonly call thinking. One way this occurs is when we process sensory data, using our memories to relate new signals from the environment to existing knowledge. Another way this occurs is when we use only our memories to reflect, reorganize, and extend the ideas and concepts. Psychologists say that we cogitate in our working memories, bringing information from our long term memories. I think of ideas and concepts as noumenal clouds, a less formal construct than mental model (e.g., Johnson-Laird, P. [1983]). A noumenon is a mental representation for a physical object, characteristic, relationship, event, or sensory experience. Clouds are highly fluid and ephemeral–like the concepts that we think about but eventually dissipate to make room for other concepts and are slightly different each time we reform them because our overall state of consciousness, experience, and physiology continually change and affect how we think. Furthermore, I think of this working memory as noumenal space where association clouds interact. In some cases, the clouds are well defined and well practiced and in other cases, the clouds are ill defined, highly ephemeral, and dense.

Likewise, there may be a few clouds or many clouds filling the mental space (it is very hard to have a single cloud—a kind of meditative or highly concentrated state). For example, a well-known fact or concept cloud such as 'eating homemade lasagna makes me happy,' coming together with a real time or past event cloud such as 'we are having lasagna tonight' can fill one's noumenal space with two easily related, well defined clouds that forms a new kind of information in the head and causes one to attain a mental state that we might call 'happy.' The new information in my head is very familiar, well defined, and rather simple. This information state may have physiological effects such as rushing to get home, or if one is very hungry, salivation.

An example of a more complex noumenal state at a more detailed level of analysis is trying to understand a new idea while reading a text passage. The information in the artifact is represented by words on the page and my sensory system detects the symbols, modulates them into signals, and transmits these signals to my brain. At physiological levels, neurons are activated at various levels, presumably based on similarity matches, and those at high enough thresholds pass on the signals. The net effect of the activated neurons is a noumenal cloud that with mental effort (rehearsal energy), can be sustained. In the case of a familiar word or phrase, the signals propagate quickly along well-connected pathways, and a well-defined noumenal cloud is formed as a mental image or idea. In the case of a new word or notion, there may not be enough neurons achieving threshold to form a well defined cloud (either because there is not enough signal getting through or because there are so many weak signals getting through that no cloud can precipitate). In this case, the signals may be reprocessed, perhaps adjusting threshold levels to allow different propagation patterns to occur. Eventually, one or more noumenal clouds precipitate from the memory traces activated and this new cloud becomes part of my mental history (we can say that one has learned the new concept) and may precipitate related clouds in the future when similar memory traces are activated. Clearly, the actual process of activating information in the head is more complex, and we now turn to more detailed distinctions of information as thought and memory.

2.2 WHAT WE KNOW

Information as what we know depends on our memory and our perception. Memory defines our past thought, sensations, and experiences and perception defines our instantaneous attention and awareness of the world. Information in this sense is what is in our memory and what our perceptual system brings to thought, which, in turn, creates memories.

2.2.1 MEMORY

Memory has been investigated by philosophers and psychologists, among others, throughout human history. Cognitive psychologists distinguish long-term memory (LTM) and short-term or activated memory. Any sensation, event, label, or experience we can recall comes from long-term memory and may be called information in the Voice1 sense. Recognizing your face as we meet on the street and saying your name in greeting are actions taken as a result of information activated in long-term memory. The capacity and duration of long-term memory is hypothesized to be infinite within the bounds of human lives—nobody has reported an 'memory full' error message that disallows any new memories to form and people 100 years of age can recall events from their childhood. Clearly, there are physiological conditions (e.g., aphasia, tumors, dementia) and psychological traumas that cause exceptions, but, generally, the amount of information our long term memories can contain is unbounded.

Psychological experiments and our everyday experience does demonstrate, however, that we cannot bring all the information in our LTM to use at any given instant. Thus, cognitive psychologists say that there is an active memory where the information from LTM can be used by conscious thought, and this is often referred to as short term memory or working memory (WM). WM is strictly limited. George Miller's famous paper in 1956 'The magic number seven plus or minus two' presents these bounds clearly and discusses the concept of information 'chunks' (like noumenal clouds) that can be processed in WM. WM not only uses information recalled from LTM, but also from the perceptual system, considered below. Miller's work and the work of Newell and Simon [1972] defined the information processing model of cognition that serves as the basis for many contemporary theories of human intellectual activity. The notion that information is held in LTM (what we know) and processed in WM as part of what we call cognition makes information the 'stuff of thought.'

Engelbart, D. [1963] and others suggest that electronic technologies can augment our memories (among other cognitive functions) and ubiquitous web access is sometimes considered to be an externalized memory function. How often do we want to do a quick lookup in a search engine during a conversation? This state has led to research on forgetting as an information process. For example, Mayer-Schönberger, V. [2009] argues that we must develop new kinds of forgetting strategies given the persistence of incidental and ephemeral information that arises in everyday life that is augmented by ubiquitous and diverse electronic technologies.

2.2.2 PERCEPTION

Although thought and memory may be strictly self-contained (reflective thought), our lives depend on ongoing interactions with the world and with diverse streams of signals that assault our perceptual system. In ordinary language, we often say that we get information from the external world through our senses. Light streaming through our eyes activates sensor cells in our retinas that, in turn, are propagated along nerve cells to our WM that activates and compares LTM images (recognition), computes (decides) whether subsequent motor or mental action is warranted, and if so, executes action (e.g., remember, move the head). Where is the information in this typical human experience that takes place in less than a second? A strong Voice1 (mental) interpretation would argue that signals and data flow from the world to our brains and only upon WM deciding whether to remember or use the data does it become information. Voice2 (communication) would call the incoming stream information.

Our perceptual systems have both voluntary and autonomic components and a large range of sensors that respond to internal (e.g., body temperature, muscle contraction) and external (e.g., light, sound, motion) stimuli. At a biological level, the stimuli are the information from the external world that is then propagated (through a variety of electro-chemical transformations along biological pathways) to cause organism action. At the social science level, the results of the stimuli change our mental state, and it is this change that defines the information received from the external world.

Regardless of whether we view the information processed by the perceptual system from biological or social perspectives, it is clear that all the stimuli from the external world are not equally important. Biologically, this notion of importance is what drives survival. Socially, this notion of importance is what we call meaningful[5]. Therefore, it is essential that humans have sophisticated information filters that limit information flows at micro (sensor) and macro (mental) levels. Our bodies and our minds learn what to ignore. Given the amount of potential information that our environment rains on us, it is likely that more effort is given to filtering than to thought and memory[6]. Socially, we exercise selective attention to the world around us. The well known 'cocktail party effect' predicts that we will turn attention to the sound of our name spoken across the room while ignoring all the other conversational streams that might be audible to us. One of the key challenges of information science is the development of models for information filtering and techniques or systems that help people manage information overload.

2.3 HOW WE KNOW

Thus far, we have considered information in the head as general biological and social phenomena. Here we consider how memory, thought, and perception work with respect to information. First, we look at the cognitive architectures that have been proposed based on introspection, observation of behavior, and interviews. Secondly, we look at how empirical studies of physiology are being combined with the different theoretical models to explain how information is processed in vivo.

2.3.1 INFORMATION PROCESSING AND COGNITIVE ARCHITECTURES

Over the past century, cognitive psychologists have posed various models of how people acquire, store, and process information. One view is that information is structured into common scenarios that people can quickly apply to real-time situations to guide behavior. Piaget posed scheme theory in the early part of the 20th century and different elaborations with computational components have been proposed (e.g., Anderson's ACT, Schank's scripts, Johnson-Laird's mental models). The main idea is that our memory and understandings of the world are determined by experiences that are classified into a small number of archetypes. Information is the stores of scenarios that are instantiated with current sensory information to make sense and take action. The development of computers in the mid twentieth century provided both new tools and models for information processing theories of thought and memory.

[5]The information retrieval community would call this relevance.

[6]Raichle and Mintun [2006] in their review of brain activity via radiologic imaging conclude that far more energy in the brain is given to internal processing of consciousness rather than to task related activity (including mental and motor exertion). It is possible that some of this internal processing is given to internal filtering.

Other models take a more granular view of information as concepts that are organized into a network of memory units (stored or active) that are connected by relationships. Quillian, R. [1967] semantic networks demonstrated how thought might progress through what he called 'spreading activation' by following relationships (links) among information units (concepts). Hutchins, E. [1995] distributed cognition extends the network organization to emphasize social relationships among the nodes. This kind of organization of information is adopted beyond human processing to realize the semantic web (Berners-Lee et al. [2001]). The key sense of information is that it is organized into a network where relationships (links) are as important as the nodes. It is the activations of the network that define information as energy.

Over time, cognitive scientists have integrated conceptual and computational models to produce architectural models that support prediction for well-defined behaviors. John Anderson's ACT-R model has evolved over decades to explain cognition as the interaction of sets of information processing modules that each accomplish different subtasks (e.g., visual recognition, rule application, motor control, task monitoring). In recent work, this model is validated with empirical brain imaging data (Anderson et al. [2008]) that begins to bridge cognitive architecture theory to human psychology.

2.3.2 BIOLOGICAL INFORMATION PROCESSING

Cognitive scientists have long sought ways to link the philosophy of mental activity to the physiology of human brain activity (e.g., Sejnowski Smith-Churchland [1989]) and much of the current work in cognitive science draws as heavily from neuroscience as it does from computer science and linguistics. Using biometric sensors (e.g., electroencephographs, eye trackers) and radiology (e.g., positron emission tomography (PET) and functional magnetic resonance imaging (fMRI)), scientists study brain activity under different information processing conditions. Eye-tracking has long been used as a way to study information processing activities such as reading (e.g., Rayner, K. [1998]) and is commonly used today by companies and scholars to study search behavior on the web.

Radiology is particularly interesting because advances in technology and lower costs have made it possible to apply it to psychological research. Neurons in the central nervous system need instant oxygen to fire and PET or fMRI imaging can measure oxygen usage in the brain. Interpretation of biometric and radiologic readings depend on a series of assumptions about what kinds of information processing leads to the readings. For example, radiologic readings measure blood flow in the brain, which is assumed to increase with increased information processing because such processing requires oxygen (e.g., see Raichle and Mintun [2006] for a review of how brain imaging applies to mental activity). Neural cells require oxygen (cell respiration) to generate Adenosine triphosphate (ATP), which is the nucleotide that effects intercellular energy transfer and intracellular metabolism. The Voice4 perspective might argue that ATP is information (energy) and monitoring blood flow via radiology traces the flow of information in the brain.

Modern cognitive science and the tools of computers and biometics have made these different approaches to explaining the how of information as thought and memory more compatible. We are a long way from unified theory of information as thought and memory that applies at molecular, cellular, and organism levels, although Friston, K. [2009] aims to do so using a model of life that has as its goal order-maximization (minimize entropy) at all levels of organization (molecule, cell, and organism).

2.4 WHAT WE FEEL

Mental states are more than activated concepts and relationships—they include physical and emotional components. Furthermore, mental states are continually changing. As Voice4 (energy) argues, information is one of the energy forms that cause these changes. Other energies also affect our mental state: various physical and spiritual forces and sensations from inside or outside the body combine with information forces to define any instantaneous mental state. Our mental state is thus determined by the environment, the body which has internal conditions related to electro-chemical balances as well as connections through the perceptual system to the environment and to other people, and the spirit which remains an intangible part of the mix. Johnson, M. [1987] has written extensively about the embodied mind and Clark, A. [1997] provides a theory of embodied mind that works in concert with the environment where information serves as the link between the mental and physical worlds. In this sense, information is only one element of the total mental state and Voice1 (mental) is overly general. One goal of meditation is to achieve a mental steady state—the absence of information and other energy forces.

We have no extensive theory of information and emotion, although the work in pharmacology and mental health that helps people manage moods and behavior is likely to spawn new developments just as advances in medical imaging to treat physical conditions spawned applications in psychological and social research. Hall, S. [2010] review of the research related to wisdom gives some direction here as wisdom is viewed as a well-functioning emotional thermostat by some theorists. Thus, wise people are able to manage emotional responses and navigate difficulty human-human interactions without resorting to systematic cognitive reasoning about the situation and subsequent emotional suppression. In essence, they just are able to easily ignore information arising in social context. As more neuroscientists consider how electro-chemical states of the brain affect mood and behavior, we will likely see new theories and explanations of how we feel and how this relates to information processing.

CHAPTER 3

Information as Communication Process

According to the *Oxford English Dictionary*, one of the original senses of the term information in 14th Century Anglo-Norman language referred to the act of providing evidence about a person. The act of informing is thus a particular kind of communication act and over time the term use broadened with respect to the substance of the informing act (e.g., oral or written words) to all kinds of purposes (e.g., teach, advocate). This sense is even broader today to include atomic and biological signaling. For our purposes here, we limit communication and the act of informing as a strictly human process.

Most people are familiar with thinking about communication and the information flow that communication enables from the perspective of who are the participants and what channels are used. From a participation perspective, we can distinguish interpersonal communication where one person informs another person (or a small group), usually with some expectation of feedback, from mass communication where one or a few persons inform large numbers of people with little expectation of feedback. In both cases, there are expectations of change in the receiver; otherwise, the communication would be ineffective. Some other special kinds of communication are scholarly communication where one or more people in a community of interest inform each other about the domain of interest through formal channels. Studies of this process often examine evidence in artifacts such as citations to infer relationships among individuals, groups, and communities (e.g., co-citations). The study of scholarly communication patterns has been termed bibliometrics, informatics, and more recently cybermetrics or webometrics[7]. The basic relationships of citations are applied to hyperlinks in the WWW and serve as one important basis for search engine algorithms.

An increasingly important kind of communication is driven by the emergence of social media in the WWW where people from all walks of life form social networks either consciously or incidentally to collaborate, commiserate, and take collective action. Wikis, blogs, and collaborative groupware are used to support collaboration on general topics or within specific communities of interest. Wikipedia is the most discussed example of collaborative sharing of knowledge. The Sloan Digital Sky Survey is an example of citizens science that involves astronomical contributions and exchange among novices and experts all over the world. The GenBank is an example of a collective effort of scientists around the world to contribute and share gene sequences and associated protein data. More general communication within social networks is supported by services such as Facebook.

[7] See Borgman and Furner [2002] for an overview of bibliometrics and exemplary studies; Cronin, B. [1995] for consideration of the motivations and uses of citations; Bjorneborn and Ingwersen [2004] and Thelwall, M. [2009] lecture in this series for treatments of webometrics.

These social networks are accelerated by computational agents that identify and highlight new kinds of possible relationships (e.g., friends of friends) and give rise to new kinds of social processes.

Design is another form of communication where artifacts are carefully organized to inform people. Design often makes innovative uses of channels and in some cases creates novel channels. Much of the work on information design is termed information architecture[8], which has become an important subfield of information science. Communication is a broad field well beyond the scope of this lecture. Linguists, anthropologists, and communication theorists have established vast literatures on communication from their respective disciplinary perspectives. In this chapter, we focus on the act of informing as a specific communication process.

3.1 HUMAN ACTS OF INFORMATION

If we were solitary beings, we would not need communication and would only be concerned with information in our heads–gathering signals from the environment and reflecting on what we know. There would be little need to create information artifacts except as personal reminders and no need to inform others about anything except perhaps to stay away. As anthropologists have demonstrated, humans are not solitary beings but highly social beings who seek and need others. The psychologist, Maslow, A. [1943] described a hierarchy of human needs that builds from the most basic physiological (food, warmth, shelter) to safety (physical and psychological) to love (need for strong social relationships), to recognition (need for both personal sense of accomplishment and broader social relationships), to self-actualization. In all of these needs, communication and information sharing play important roles. Thus, humans require rich and robust means to interact with other people to build and maintain relationships in order to satisfy basic social needs in their lives.

The means of interaction are information acts. These information acts range from physical acts such as eye contact, gesture, touch, and embrace, to culturally developed acts such as speaking, writing, and singing. Additionally, we have developed a variety of tools and techniques that amplify and augment our information acts beyond the constraints of space and time. A megaphone amplifies voice, musical instruments augment our emotional and intellectual expressions, and recording devices ranging from sculpture and the written word to cameras and telecommunications systems increase the spatial range and make our information acts persistent over time. These persistent acts themselves are information artifacts (ala Voice3) that give rise to entire industries, and as considered in the next chapter, they have become the dominant sense of information in the 21st century.

Actions do not stand alone. To understand an action, we must consider at least three things: the actor, the act itself, and the object of the action. Basic human information acts are defined by a sequence of three components: the actor's intention, execution, and effect. These components may be repeated within a single sequence, and the sequence is often repeated over time in human-human communication. We examine each of these components in turn.

[8] See Rosenfeld and Morville [2002] for an introductory text on information architecture, and the Ding and Lin [2009] lecture in this series.

3.1.1 INTENTION

Information actors may intend to change an external entity, a social status, or a personal state. The act of informing is often driven by an intention to change a state of the world, most typically mental state in one or more other humans who may, in turn, act on the physical world. Actors may act alone or in groups and perform information acts with a variety of intentions ranging from altruistic to masochistic We may intend to help someone else solve a problem, avoid danger, or achieve a goal; or to adopt an attitude or belief; or take some action.

The intent of the action may be directed at one or many. Thus, information acts may involve one-to-one, one-to-many, many-to-one, or many-to-many human actors. Most of our information acts are one-to-one as we explain, cajole, and discuss ideas and decisions with others. Media broadcasts illustrate the one-to-many (few-to-many more typically) cases where one person (or group) aims to change the mental states of many. Grass-roots movements where many people inform leaders about their positions on legal or other issues illustrates the many-to-one case. Electronic technologies such as social media offer dramatic effects in this regard. For example, the U.S. government responded to the large influx of emails and other electronic comments to different regulations by creating the regulation.gov website that in 2010 listed more than 8000 regulations and invites comments from the public. Many-to-many information acts are illustrated by political movements where one group aims to influence large numbers of the public. Today's social media support crowd-sourcing and affinity groups that intend to affect groups large and small.

Another intention factor is whether the actor(s) expect to receive feedback. In one-to-one informing, we typically expect at least acknowledgement of understanding and perhaps response. Many mass media campaigns do not expect explicit feedback but do expect change in behavior or mental state. It is interesting to see how traditional mass media like newspapers, radio, and television have begun to encourage more feedback mechanisms to try and maintain market share in the face of more interactive many-to-many venues on the WWW.

In addition to expectation of feedback, people expect to be able to decide whether to participate in communication. Whitworth and Whitworth [2004] offer an analysis of email spam as a side effect of poor technical infrastructure design that ignores the social constructs of human communication. In addition to taking advantage of the capability to copy electronic items with little cost, spammers leverage the fact that email systems do not require return-to-sender functions, and this violates the natural social contract of human communication that requires choice to participate.

Not all information acts are intended to change others. Sometimes, the intention is simply self-gratification. For example, the need to express oneself regardless of audience can be the intent that drives an information act. Social media such as blogs, Facebook, or Twitter provide forums for such self-expression. We all know people who not only like to hear themselves talk but actually need to do so to maintain their sense of identity. The motivation for acts of information not directed at others can range from madness to artistic expression.

More typically, the intention is to build and maintain relationships for the purpose of social bonding—keeping channels open is the main intention rather than causing effect in the receiving person. Marshall and Bly [2004] studied how people share electronic and physical clippings and showed that people sometimes did so with the intent to build and maintain social relationships. The content was not pertinent but rather demonstrates information as social act. Social exchange theory [Homans, G., 1958] posits that people perform information acts with specific psychological motivations and expectations of reciprocity. Thibaut and Kelley [1959] delineate expected reciprocity, altruism, direct reward, and reputation as intentions for social exchange. Brown and Duguid [2000] consider the roles that software agents have begun to play as actors in socially-situated information acts and the enormous popularity of social media (e.g., blogs, Facebook, Twitter) demonstrate the power of social intention in information acts.

3.1.2 EXECUTION

Information acts consist of patterns of cues that have cultural significance. The patterns are variations, repetitions, and juxtapositions and the cues can be shapes, pitches, colors, motions, or other discrete physical conditions. The performance of information acts depend on several factors: the representation scheme or language used, the medium or channel used, the time interval over which the act takes place, and the skill of the actor. Throughout history, people have created a variety of tools and techniques to aid them in executing information acts and considerable portions of our educational systems are devoted to developing skills to use these tools and techniques. For example, the representation scheme of human language expressed as ink on paper has proven useful enough to cause us to devote years of our lives to developing skills to execute and interpret information acts that use this scheme and medium. Changing the medium from physical analog to electronic digital changes some of the skills required to execute information acts and the scale of distribution and impact. These changes ultimately influence intentionality as people adjust to the new performance requirements and scale of impact. For example, Stutzman and Kramer-Duffield [2010] studied how college students changed privacy settings in Facebook over time as the scope of profile propagation became more apparent to them.

From a channel perspective, humans use a variety of naturally occurring channels and have developed a host of artificial channels to inform others. Likewise, we have evolved a number of representation schemes to take advantage of these channels. Pressure waves created by vocal chords, propagated through space, and decoded by the auditory system allow a variety of human spoken languages for face-to-face communication. Informal face-to-face communication was one of the earliest and is the most basic and common way that people inform others and become informed by others. Human social organization led to more formalized schemes for using oral language such as rhetoric, debate, and politics. Oral and non-verbal accompaniments are ephemeral in that the act is fixed in duration and persists only in human memory. The Homeric oral tradition of story telling as informative act demands substantial practice and skill.

Early humans also leveraged elements in the environment to produce artifacts that inform others over time. Cave paintings, rock formations, and wood carvings have been used for eons to inform. The Inuit carved totem poles to tell stories and inform descendants about a family or tribe (note that these were not public but highly personalized to the family or tribe). Aboriginal peoples painted or carved artifacts to inform others about how to hunt, cook, and celebrate. Painting and sculpture serve as persistent information acts and highly skilled creators of such artifacts stand as historical figures in most cultures.

Record keeping symbols and pictorial artifacts led to more complex symbolic codings for information. A crucially important development in the evolution of informative acts was the invention of written languages linked to oral languages. Written language depends on culturally agreed upon sets of symbols that actors perform with some tool (e.g., charcoal) upon a variety of physical channels (substrates) ranging from papyrus to neon signs. Bolter, J. [1991] describes hypertext as the technological extension of written language as thought.

A recent revolutionary development for information acts that continues to transform human society is the set of different electronic channels of communication. The telegraph, radio, the telephone, television, and today, the WWW have dramatically augmented human culture and our capabilities to inform[9]. Just as information as artifact has been dramatically affected by electronic technologies, so information as communication process has been massively affected. The telephone and email have revolutionized our capabilities to inform others, and television and the World Wide Web have added entirely new dimensions to mass information acts and scholarly communication. The concept of literacy has been expanded to include skills with new kinds of tools and techniques for executing information acts (e.g., Eisenberg, B. [2008]).

Today, humans have a wide range of representation schemes and associated channels with which to execute information acts. We invest substantial time and effort to acquire and practice skills to apply these schemes and channels to execute information acts for a variety of intentions. These acts may take instants or years to perform and may be ephemeral (e.g., speech, dance, theater, music) or persist over time due to the properties of the scheme and channel or the use of recording technologies. This richness of human investment reflects the many kinds of effects information acts have on human life. We spend countless hours learning to execute these acts and our economic well-being is often directly influenced by our abilities to learn and execute information acts.

3.1.3 THE EFFECTS OF INFORMATION ACTS

Information acts may have direct or incidental effects. The direct effects are changes in belief or behavior and may meet the intentions of the actor who performed the act or chose to not perform it. Negative campaign ads in political races are used because they are effective in changing or forming beliefs in voter's minds (and thus subsequent voting behavior). However, in some cases, they can backfire and lead to the opposite effects. Groups may try to minimize the backfire effects by disguising the actual actor through veneers of surrogate groups or entities that execute the information act.

As noted in the sections above, most information acts are intended to create direct effects. Informing children about the beauties and dangers of life is not always effective. The effects are

[9]See the Victorian Internet by Standage, T. [1998] for a fascinating set of parallels between the development and impact of the telegraph and the Internet.

important and effective often enough to make such information acts routine for parents and teachers. Leaders in all walks of life use information acts to direct the action of others. Thus, enormous time and effort is given to learning how to execute information acts precisely because the direct effects are so powerful.

Information acts often lead to incidental effects in several ways. Foremost, they are manifested in artifacts or spawn associated artifacts. The book is a manifestation of a substantial informative act and itself becomes an object of interest. The recording of a radio broadcast or a series of blog postings discussing a speech or a review of a movie or book are all derivative artifactual effects that accrue based on the primary information act. Additionally, the direct effects of information acts recursively spawn new kinds of incidental effects. Much of the information industry, including library and communication studies, has evolved as a result of the desire to understand and preserve information acts and the subsequent acts and artifacts that they spawn.

Take, for example, the layers of new services that arise from information acts. Publishers capture important acts and distribute them. Libraries and archives collect the manifestations of these acts, preserve them for future use, and make them available to people. Reference services and search engines exist to help people locate and access manifestations of the acts and associated acts that provide context for them. Within a library, reference interviews may be conducted to help the information seeker articulate their need and the library agent to understand this need. Thus, a series of layers of services have evolved over time to make certain that the most important information acts go beyond their ephemeral states to serve the long-term needs of the culture. This elaborate and expensive enterprise illustrates the importance that information acts play in human existence.

3.2 INTERACTION AS INFORMATION ACT

People interact with other people in a rich variety of ways. The traditional way that information enters interpersonal interactions is as an intermediary or medium of interaction. Thus, a conversational interaction between two people is mediated through the utterances each makes and the protocols for structuring the utterances (e.g., turn taking, openings and closings). Human-human interaction is more specific than human-human communication in that interaction requires one or more feedback cycles and commonly takes several cycles. Mass communication and scholarly communication have traditionally been much less interactive than interpersonal communication. Human-human interaction is of fundamental interest to all the social sciences, and there are rich literatures that focus on the actors (e.g., psychology), the actions (e.g., linguistics, communication studies), and the contexts of interaction (e.g., sociology, economics). Human-human interaction is dependent on information as a process, thus there are strong overlaps between communication theory and information theory.

In thousands of years of human experience, we have learned to use many kinds of tools, but the preeminent kind of interaction has been human-human interaction. Because human-human communication and interaction are so fundamental to individual life and the development of social institutions, people have created tools to facilitate these activities. There is substantial literature on technology-mediated communication in its own right. From the time of Socrates', concerns about writing as a mode of expression that would linearize thought, the development of new technologies to foster communication and interaction have generated interest and debate.

Mediated communication studies were spurred by the technical developments of the nineteen and twentieth centuries and especially by the advent of electronic technologies such as radio, television, and networked computing. Marshall McCluhan (1964) generated substantial debate when he claimed that the technical medium strongly influences the meaning constructed by the receiver of a message. More contemporary scholars put attention on the technology characteristics that facilitate communication and collaboration in work environments. Reeves and Nass [1996] go so far as to argue that people do and must treat media as they treat other people because our evolution has been based primarily on interactions with other people. There is a substantial field known as computer-supported collaborative work that considers how electronic technology can be leveraged for communication and collaboration purposes[10]. A more mature and broader field with forty years of development is called human-computer interaction (HCI), which focuses attention on why and how people interact with the technology itself with an eye toward how the technology can be improved to better assist human performance[11]. For example, graphical user interfaces, menus, and clickable hyperlinks are much easier for novices and casual users than are command languages. As the technologies themselves advance, they begin to fade into the background—become part of the infrastructure—and thus allow researchers and practitioners to focus more closely on the actor intentions and the results of interaction. It is in this vein that information-interaction has begun to emerge as a research front[12].

Therefore, changes in the nature of information due to electronic technologies are enabling substantial *human-information interactions*. Malleability and conditionality made possible through programming, instant global communication linkages, and practical, cheap mass storage unite to make information in electronic form a first class interactor in human-information interactions. By first class, we mean able to initiate interactions and engage in multiple cycles of feedback. The fact that information objects can present many fixed or randomly generated states gives them a kind of dynamism that has only been exhibited by people and to a lesser extent by animals throughout human history. This creates new opportunities as well as challenges. It enriches and complicates our lives and the information landscape. These developments are evolutionary rather than revolutionary—they may displace or mitigate some kinds of entities and activities, but they mainly amplify and augment the range and fidelity of entities and activities and lead to information as experience—a self or social informing act. Human information interaction is treated more fully in Chapter 6.

In this chapter, information as action taken by humans to effect changes in the world or satisfy psychological and social needs was delineated across three stages of activity: intention, execution, and effects. Over time, people created a range of tools and techniques to execute information acts

[10]There is an annual CSCW conference and several journals devoted to collaborative systems and processes.
[11]CHI is one of the largest conferences held by the ACM and there are many books and journals devoted to it.
[12]Mark Weiser popularized the concept of ubiquitous computing and later calm computing to argue how the technology fades to the background. The European Union has funded the disappearing computer initiative (`http://www.disappearing-computer.net/`).

of different complexities and persistencies. The effectiveness of information acts has led to massive enterprises and human investments across many layers of human endeavor. The products of the acts persist as a range of artifacts that have become know as information, a sense that we turn to in the next chapter.

CHAPTER 4

Information as Artifact

The most common sense of information in popular culture considers information to be the physical objects that are created to express ideas and meaning. Objects such as newspapers, books, and television and radio streams are said to be both informative (Voice2) as well as to be information objects themselves. In this chapter, this physical sense of information as object that carries meaning is considered from a human-centered perspective with emphasis on how electronic digital artifacts are augmenting the many physical information artifacts that have influenced cultural and economic development over time.

There is something comforting in thinking about information as a tangible object—a book or a DVD is well specifiable, it can be put in a physical space, quickly identified with unaided human senses, and remains stable in form over time so that people may anchor their internal sense about the information in their minds. It is information as artifact that has been most obviously affected by electronic media because the crafting, managing, and representing functions are strongly affected. Different types of artifacts have more readily migrated to electronic forms than others. For example, reference materials, manuals, and other materials used in on-demand situations in small portions have quickly migrated to electronic form whereas holistic objects such as novels are doing so more slowly. It is likely that physical and electronic information artifacts will continue to co-exist for a variety of reasons.

Books are perhaps the canonical information artifact. The book has evolved over hundreds of years as a conduit for human ideas. The information in an author's mind is mapped onto a coding system (a written language) and the mapping is manifested by some physical process such as ink carefully placed on paper. The ink and paper artifact is then made available to others who map the ink bits onto their own minds as information states. This captures the three senses of information—information in the head is manifested as an artifact, which, in turn, informs other people about the original ideas in the mind of the author. The artifact itself can be simple and highly functional, such as a lightweight paperback, or stunningly beautiful such as a coffee table book. Some scholars distinguish the ideas in the author's head as the 'work' from the book as an artifact that manifests or approximates those ideas. Rather than the physical ink and paper[13], the ideas represented (the work) are what most people actually mean when they talk about the 'information in books.' In popular parlance, a library with two million books is said to contain more information than a library with exactly one million of those books. This is a nice property of information artifacts since we

[13]The library community has begun to define and adopt Functional Requirements for Bibliographic Records (FRBR) that distinguish four levels of representation in the bibliographic metadata that accompany all forms of information artifacts (work, expression, manifestation, and item).

can measure the volume of information by counting physical attributes, whereas it is very hard to measure the value of mental states.

Thus, books are well known information artifacts that have various properties and meanings for people. Although several electronic alternatives have been created for physical books, paper book production and sales continue to increase in spite of most other forms of information shifting to electronic forms. Audio books and electronic books (e-books) have generated considerable interest and investment and are generating strongly increasing sales revenue as new kinds of readers become popular[14]. Nonetheless, paper books remain highly viable. Some of this is due to the convenience of the artifact (easy to handle, highly portable and flexible, excellent visual properties for human eyes, inexpensive) and some are due to corresponding limitations of electronic books (e.g., power requirements, visual limitations in strong light)[15]. Given that reference books, manuals, and other books that are not meant to be read in full or in linear fashion have become largely electronic, it seems that the physical properties of books that aim to sustain a reader's attention for many hours are important.

Photographs and film are powerful information artifacts that are have been revolutionized by electronic technology. Consumer grade cameras in the first part of the twentieth century and video cameras in the last part of the century democratized photography and video production by allowing non experts to capture images and events on film or video. Digital cameras and easy to use editing and management software add another qualitative shift that empowers people to represent moments of reality easily and cheaply, add additional value to these representations, and manage large collections of personal images or videos over time. Ease of capture and low cost empowers people to experiment, capture more variations, and gain much more experience with the technology and the resulting information artifacts than is possible with film-based techniques. Editing and management software allows people to alter the raw captures (ranging from removing the red eye effects from a flash to artistic effects with overlays and pixel-level changes) and to add descriptive metadata to help find the artifacts at a later time.

All these advantages add up to new capabilities and applications for still and motion images. Furthermore, another qualitative shift in empowerment comes when broadband and wireless networking are added to the mix. Emailing digital photographs or posting them to file sharing sites, and peer-to-peer web cameras allow travelers, soldiers, and distant relatives to stay in touch with their families from afar. The integration of the camera and telephone offers new kinds of communication in which images are as natural to use in face-to-face conversation as recalled memories. Likewise, distributed conversations with high fidelity are possible as cell phones integrate multiple cameras that support video conferencing. These trends in image and video creation, sharing, and viewing have significant impact on Internet growth and use. For example, in 2010 Yahoo's image server used

[14]An August 2010 news release for Random House noted that 10% of sales were for e-books. `http://www.reuters.com/article/idUSTRE6700Z620100801`

[15]See Dominick, J. [2004] for an extensive and longitudinal study of e-books in university courses that demonstrate the limitations of e-books and the complex set of contextual factors that influence book usage in learning.

2.9% of Internet bandwidth and YouTube used 10.2%, both of which exceeded the 2.5% used by the Google search engine (Leavitt, N. [2010]).

Similar examples can be traced for music and voice recordings and other human information interactions. As with other information artifacts, electronic technologies blur distinctions between the information in a person's head (Voice1), the process of sharing memories and ideas (Voice2), and the artifacts that facilitate thinking and communication (Voice3). Information as artifact is especially important from a practical perspective because it gives rise to the entire information industry that includes publishing, records management, and libraries. It also facilitates engineering enterprises such as database management and information retrieval that support the artifacts.

4.1 ARTIFACTS

Buckland, M. [1991a] provides a classical treatment of information as 'thing' exploring the history of functional views of 'document' objects that depend on materiality, intentionality, crafted representation, and human recognition rather than specific format (Buckland, M. [1997]). In this lecture, I aim to broaden this notion to multimedia and prefer to highlight the distinction by using the term 'artifact' rather than document. The general term artifact includes a broad range of objects while retaining the connotation that the information object is created by human effort. This is an important qualification because it omits a variety of signals and data that occur in nature and are often called information. For example, sequences of nucleotides in DNA and RNA are naturally occurring data that are critical to life but are not artifacts but rather naturally occurring biofacts[16]. The diagrams, mass spectrometry displays, and microarray outputs of gene or protein biofacts, however, are artifacts because they are representations created by humans.

Artifacts may be anticipated or not (e.g., in software engineering and medicine, the term artifact often refers to unintended consequences of algorithmic or biological processes). Here we are interested in objects that are created purposefully by people and with a variety of tools. The purposes and tools, in turn, strongly influence the qualities and uses of information artifacts.

Two important factors define information artifacts: form and life-cycle. These characteristics are roughly concerned with 'what' and 'how' qualities: what comprises an information artifact and how it is created, managed, shared, and used.

4.2 FORM

Three components interact to determine artifact form: physical substrate, method for changing substrate, and tool. The physical substrate serves as a ground upon which humans act with some body extension (tool) to alter the substrate according to some set of rules. The resulting object is information as thing. Figure 4.1 illustrates these components and lists some examples for paper-based objects such as a handwritten letter, stone-based objects such as sculpture, chemically treated paper-based

[16] Likewise, various biochemical and electrochemical processes that are crucial to life such as ion channel signaling and ATP-ADP exchanges might be included in the information as communication act sense, but we will limit the acts to human acts.

objects such as photographs, magnetic surface objects such as computer disks, and electromagnetic streams such as radio waves. Different combinations of substrate, tools, and structures define what are termed media. Many of the components are usable in different combinations. For example, different methods and tools can be used with paper as a substrate. Likewise, rules such as Morse code can be used with electricity, light, or sound substrates. What is particularly important in the past 60 years is that enormous effort has been given to mapping physical and analog artifacts to electronic digital artifacts to take advantage of storage/copy/share/manage properties.

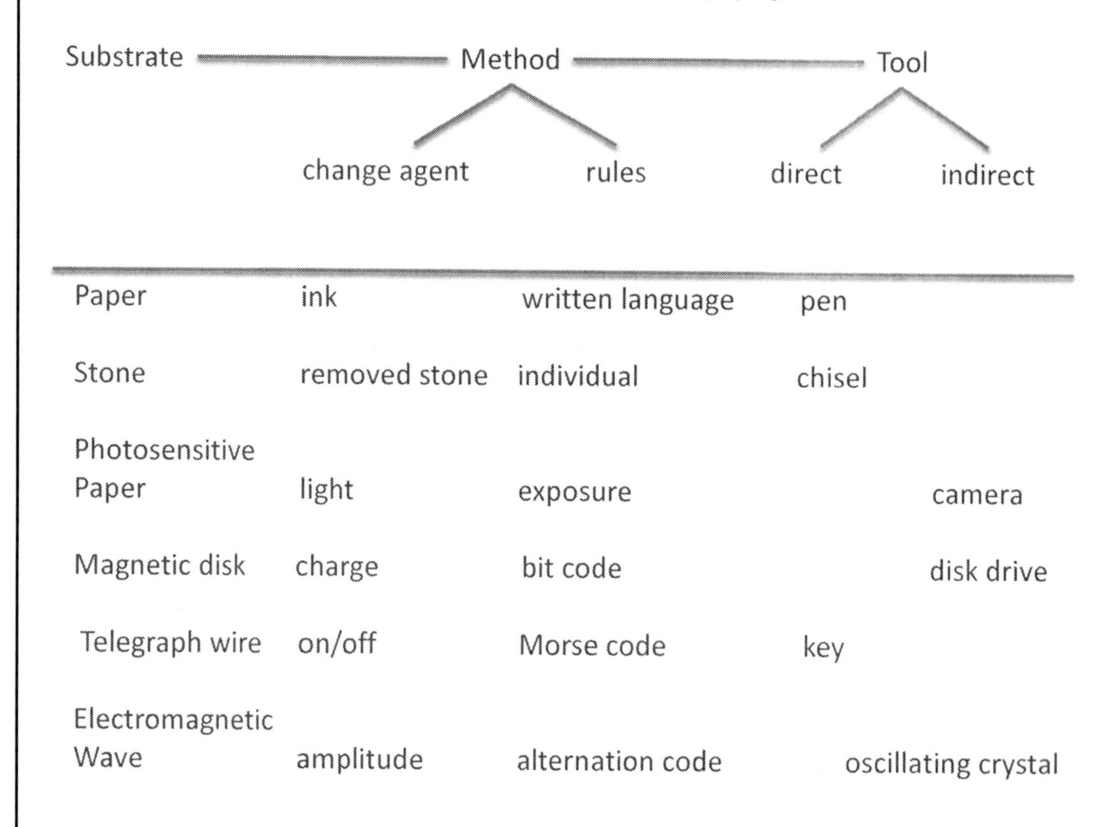

Figure 4.1: Information Artifact Components[17].

4.2.1 SUBSTRATE

Over time, humans have used many types of matter as substrates, ranging from naturally occurring matter such as stone, clay, wax, animal bones or hides, and plant material (palm leaves, papyrus, wood), to manufactured matter such as canvas, paper, and rope. As science and industry advanced,

[17]These specific examples have many variants. For example, today's digital cameras are battery powered rather than manual; there are many forms of electromagnetic radiation and tool variants to manipulate the ground signal.

new substrates were created such as photosensitive paper; wax cylinders; phonographs records; nitrate film; and a plethora of 20th century composite metal, plastic, and silicone. The electronic age brought electromagnetic and optical streams as new substrates for an array of ephemeral information artifacts. These substrates serve as the 'ground' for information artifacts; they are not sufficient alone but require a method and associated set of tools for altering the substrate.

The development and widespread adoption of time-based physical streams such as radio waves as substrates has revolutionized information industries and broadened the notion of information artifact beyond tangible objects. One important requirement of streaming substrates is the need for continuous power to sustain the substrate, and this has led to many companion artifacts that preserve the time-based artifact for reuse at later times. A live interview may be coded onto radio waves that propagate toward listeners, but recordings are made to more permanent substrates (e.g., tape or disk). Additionally, because the electronic substrates are ephemeral, tools and methods to represent what is stored are required for people to use the artifact. Thus, the recorded interview may exist on a tape-based artifact, but a player is required to activate it and re-generate the ephemeral audio artifact onto air pressure waves.

Rather than a self-contained artifact like a book, electronic artifacts consist of suites of artifacts specific to creation, storage, and use. Use of electronic methods and tools add new kinds of capabilities (e.g., copy, transfer, and small physical space requirements) but require multiple artifacts to be useful over time. These multiple artifacts (e.g., the ephemerally generated original signal artifact, the storage artifact, and the re-representation artifact) also complicate the management process discussed in the life spiral section below. Many electronic substrates converge in the WWW, which is the primary substrate for information in the 21st century and lead to a new form of information explicated in Chapter 6.

4.2.2 METHODS

Methods depend on two subcomponents: a change agent and a set of rules for manipulating the changes to the substrate. The change agent can be a different type of matter (e.g., paint, ink) that overlays the substrate or a modification of the substrate itself (e.g., carvings, indentations, holes, electrical charges, wave amplitude changes). The change agent is directly related to tools that are used to alter the substrate. For example, paint can be applied with fingers, brushes, knives, trowels, or compressed air sprayers.

Rules are socially agreed upon coding schemes such as human languages, musical notations, and a variety of codes created to optimize substrate-method-tool combinations. For example, written English language uses 26 characters that are organized into words and phrases that represent concepts. Learning to use (write and read) these codes is a complex process that is deemed worthy of many hours of effort by everyone who wishes to become literate. Note that there are many layers of coding that take place as concepts are mapped to text statements, words, and characters. The number of characters and variability of word length offers enormous possibilities for rich expressions (e.g., there are 26^5=11,881,376 possible 5 character 'words' in English[18]) and cultural constraints that determine which expressions map to meaning (e.g., there are about half a million words regardless of length in the English dictionary excluding scientific/technical words).

Electronic computers depend on layers of code that map human-readable codes such as words and numbers onto binary machine states (e.g., high and low voltages). ASCII code was developed to map characters onto 8 binary digits. For example, the upper case character 'A' is represented by 01000001. These 8-bit codes are called bytes and are the base unit of processing for computers[19]. Today, different computers use different numbers of bits as their fundamental unit of computation (e.g., 8, 16, 32, 40, 64 bits) and more extensive codes such as UNICODE allow many more than 128 (2^7) characters to be represented. Progress is also being made toward quantum computers that use qbits that represent probabilities of state rather than the strictly 0 or 1 states of binary digits (bits). Layers of code are used so that people can map ideas onto more human manageable codes. For example, a programming language like Java is closer to human language and thus easier for humans to use. The Java code (source code) is then mapped automatically to the operating system and eventually to machine code (object code) in the machine substrate. A particularly useful coding scheme for electronic information processing and retrieval is the Extensible Markup Language (XML) that codes the structure of books, articles, webpages, and other artifacts. XML coding supports alternative formatting/display as well as new kinds of retrieval that targets items within artifacts (see the lecture on XML retrieval by Lalmas in this series).

Coding schemes can be classified on a continuum from visceral—naturally recognizable by humans—to symbolic—requiring learning. Iconic codes for international travelers aim to provide immediate context-dependent messages, (e.g., an airplane for an airport, a cigarette with slash through it for no-smoking). The richness and diversity of human thought and the preference to focus on ideas rather than coding lead to the creation of complex coding schemes such as human language that are highly expressive and natural to human physical capabilities. The tradeoff for this richness is substantial learning costs to become facile at coding and decoding. Mathematics provides some of the most classical examples of highly symbolic codes to express common or complex relationships with precision.

4.2.3 TOOLS

Applying methods requires some kind of extension of the human body—a tool. Tools can be directly manipulated (powered) by humans or they may be powered by external forces (e.g., mechanical springs, falling water, wind, or electrical current). Direct tools include objects such as human hands, knives, brushes, and charcoal. Powered tools include tools such as lathes, digital cameras, and computers. As we discuss below, learning to use these tools is a requisite for literacy.

[18]This is higher than the actual number of 5-letter English words because it allows 'words' such as RRRRR.
[19]In ASCII, one bit in each byte is reserved for control/error checking (e.g., parity bit).

4.3 INFORMATION LIFE SPIRAL

People create information artifacts in a variety of ways and with enormous ranges of effort and intention. Writing a note to oneself as a reminder to do a task later in the day is easy, personal, and useful for only a very short time interval. Writing a book is difficult, intended to be public and useful for long time intervals. Professional work typically entails substantial effort in creating and using information artifacts that record activity, communicate intentions, and facilitate planning and execution. The pervasiveness of effort and importance these artifacts have for work and play have generated major industries devoted to information artifact creation, management, distribution, and use.

Creation, dissemination, use, and dispensation of information artifacts in the publishing and digital library sense has been described by Hodge, G. [2000] and others as the information life cycle. Figure 4.2 depicts the information life spiral. The figure depicts one whorl of a spiral. I prefer a spiral rather than a cycle to reinforce the notion that information evolves and is reused or adapted to create new information artifacts. The nautilus-like life spiral is also meant to suggest that individual information artifacts evolve and expand even within the phases of a whorl. The Public-use arrow in Figure 4.2 depicts the next whorl in the spiral as use leads to creation; however, it is useful to keep in mind that each phase may itself have iterations that modify the artifact. One important effect of the WWW substrate is that the whorl iteration of adaptation and reuse occur with very high frequency and at global scale. For example, moments after posting a YouTube video, many thousands of people around the globe can view, annotate, and reuse the video to create derivatives.

The spiral includes artifacts that are for personal use only and those that are meant for sharing beyond an individual. In the case of personal information artifacts, creation, management, and use phases may include destruction of the artifact or using it to create new information artifacts (thus the dotted line from Personal Use to create defines a subspiral of personal information). Creation may be done using different substrates, methods, and tools according to the need and skills of the creator. The creation process can be quick or take place over time. The artifact and preliminary versions must be managed during and after creation. A painting done over a week requires storage and protection each night; an essay done using an electronic word processing system must be named and versions kept (and backed up) or deleted; a reminder note to oneself must be positioned in space and time. At some point, the information artifact is used. The reminder note is consulted and an action taken, the essay is finalized and submitted for publication, and the painting is viewed.

Information artifacts are created with some sharing intention. The sharing intention ranges from completely private to semi-private (sharable with selected others) to fully public (openly published). Public collections of information artifacts must deal with challenges of collection size and diversity and thus sophisticated storage and access techniques have arisen that go beyond the requirements for managing and using personal collections. Sharing intention leads to the important issue of intellectual property and associated issues such as copyright. These issues are beyond the scope of this lecture, but see Lessig, L. [2008] and Lanier, J. [2010] for countervailing arguments on open source and restricted intellectual property.

Whether personal or public, information artifacts acquire history and are often used as the basis for new artifact creations and thus lead to new layers of the spiral. Each of the stages in the life spiral is considered, in turn, with an eye toward the changes that electronic tools and artifacts bring.

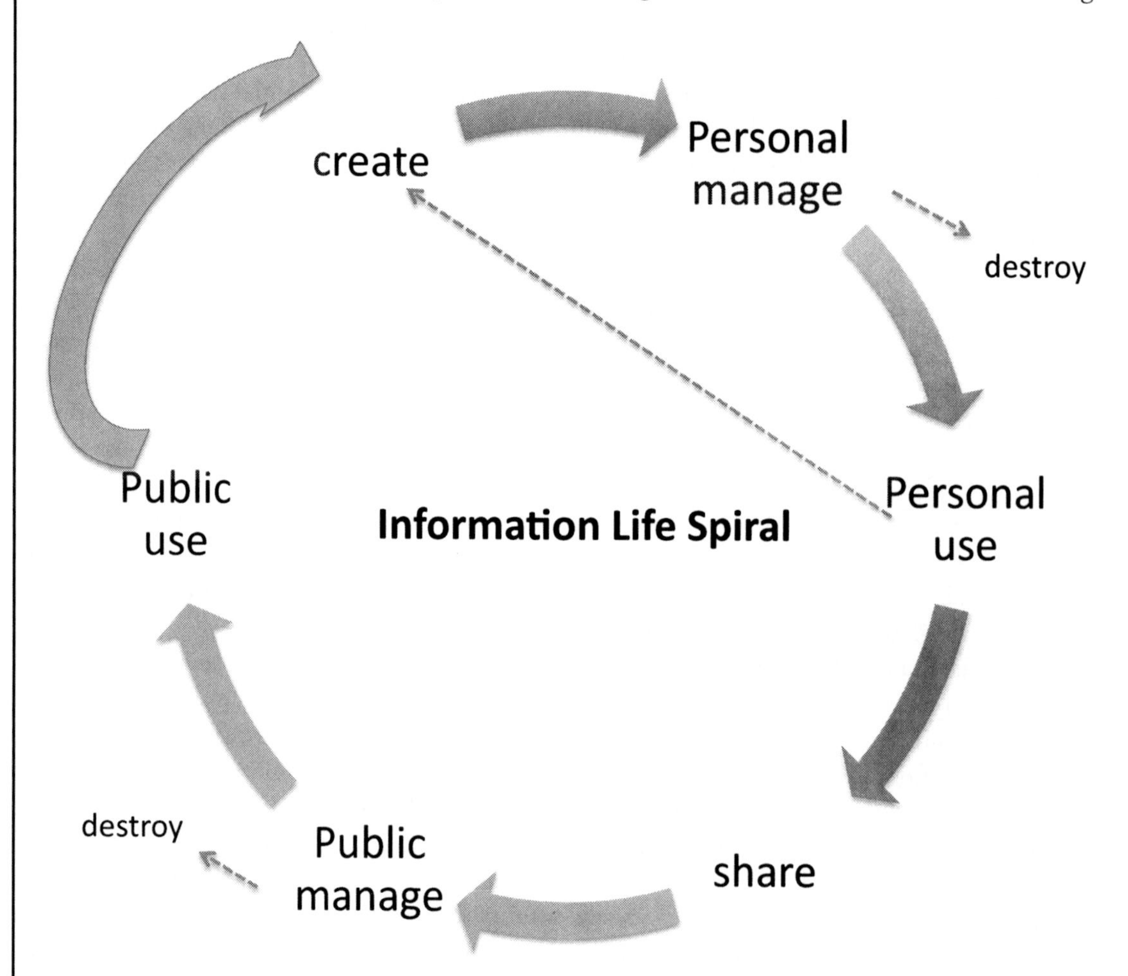

Figure 4.2: The Information Life Spiral.

4.3.1 CREATE

People create information artifacts for many different reasons. Artifacts that are meant to be completely private tend to be created as reminders (e.g., notes from a meeting) or as life records (e.g., diaries). There are extreme cases where people are recording many aspects of their lives using a variety of cameras, sensors, and recorders. Gordon Bell's MyLifeBits project is an example of saving all telephone, email, written documents, and web browsing in order to better understand how to categorize and organize our histories (Bell and Gemmell [2009]; Gemmell et al. [2006]). More radical examples are efforts to mount video cameras and various kinds of physiological, spatial, temporal, and motion sensors on human bodies (Sawahata and Aizawa [2003]). Pentland, A. [2008] discusses the many kinds of personal sensor information that people will generate and manage. He calls this reality mining and argues that people will use these data streams to improve the quality of their health and performance. Sellen and Whittaker [2010] argue for starting with the psychology of human memory and action rather than technology capabilities as designers create lifelogging applications.

More typically, people create artifacts that are meant to be shared with selected people. For example, photos are often shared with family members and friends, email messages or reports are meant to be shared with co-workers. Artifacts that are meant to be shared widely have traditionally depended on publishing industries. Musicians and authors have long depended on publishers to vet, copy and distribute their work and the publishing industries over time have gained substantial influence on work creation. Electronic tools and the emergence of the WWW infrastructure has made it possible for individual creators to distribute their artifacts without using traditional publishers and has enabled the growth of alternative publishing models (e.g., blogs allow cheap personal publishing, Wikipedia supports and depends on collaborative creation).

People create information artifacts by using tools that are closely coupled to the physical substrate and methods of altering the substrate. Use of tools such as pen, paintbrush, and piano requires skills that require substantial practice. Creators of information artifacts must therefore have two kinds of talent: creativity in leveraging method to express ideas, and skill with tools to craft the artifact. The development of electronic technologies has brought significant change to the information artifact creation process. As noted in the previous section, creation of electronic artifacts requires electric power for creation and persistence. A robust electric grid, advanced batteries, and a range of new storage substrates support continued evolution of new kinds of artifact suites.

New kinds of artifact creation tools that enable people with minimal training to create a variety of information artifacts. For example, easy-to-use and inexpensive video cameras allow almost anyone (including young children) to create video artifacts, and electronic synthesizers allow novices to mix and create sophisticated ensembles without learning the manual skills to play instruments[20]. WWW infrastructure then allows easy sharing of electronic artifacts, which, in turn, motivates reuse and even more creation. Likewise, word processing software and blog applications have allowed new ranges of citizen journalism. It is often said that electronic media and the WWW have democratized publishing. Regardless of what one thinks about the quality of the creations, it certainly has increased the volume of artifacts created and shared.

[20]Note that easy tool use is not sufficient for creation of quality artifacts. See D.J. Spooky (Miller, P. [2004]) for a remix culture perspective.

4.3.2 PERSONAL MANAGE

As people create information artifacts, they must also consider what will become of those artifacts over time. If the artifact is created as a personal memory, it must be preserved and findable. If the artifact has served its purpose, it must be discarded or archived. Managing books, magazines, phonograph records, photographs, and other artifacts requires physical space (e.g., shelves) as well as schemes for labeling and organizing collections. Titles, creation dates, descriptions, and other discrimination cues used to label individual artifacts are referred to as metadata. Organizational schemes range from simple alphabetical orderings by title to topical orderings that are idiosyncratic to individuals. The complexity of metadata and organization are dependent on the size of the collection and the proclivities of the manager.

Electronic artifacts have created new kinds of management challenges as well as opportunities. People create thousands of electronic files (text files, images, spreadsheets, email messages) and tend to adapt physical metadata and organizational schemes to manage them. Because a file does not take much physical space (many orders of magnitude less) and is easily copied, many versions can be placed in different storage bins (e.g., different file folders or different machines). More importantly, computational processes allow the indexing of each file in a multiplicity of ways, thus allowing search for the artifacts by a large range of metadata features (e.g., date, title, author, color, run-time, topic).

The physical space required to keep information artifacts such as magazines causes most individuals to recycle/destroy them after some period of use. As electronic files take up little physical space, people often adopt a 'save everything' practice, arguing that storage is cheaper than taking the time to delete files. Such practices in government and business are tempered by Freedom of Information Act request filings and accidental or subpoenaed releases of data that cause creators to be more judicious about archiving or destroying files. Most individuals have not adopted thoughtful file destruction practices. Personal information management has become an important area of research and development to address the challenges of personal artifact management and use (see Jones and Teevan [2007] and Jones, W. [in press] in this Lecture Series for treatments of PIM).

4.3.3 PERSONAL USE

Information artifacts are created to be used. There are many uses to which artifacts may be put, but common uses include the following: reminders, records of activities, decision aids, and consumption/reuse for pleasure. Beyond these typical uses, information artifacts can be used as doorstops or as indicators of wealth or erudition. Regardless of the intention of use, artifacts must be recognized and positioned (found), and in the typical cases, they are processed to extract meaning (turned into information in the head, taking the sense of information from Chapter 2).

As discussed in the components section, electronic artifacts require power to be used (perceived) by humans. The dynamic nature of these artifact suites changes the way that people use them. Today's e-book reading devices aim to emulate the perceptual qualities of paper books, but they add hyperlinks and active media (see Marshall, C. [2009] in this series for a lecture on reading and writing e-books). These artifacts will change reading behavior. Renear and Palmer [2009] provide evidence for the ways that electronic texts are changing the nature of scientific reading and information consumption. The evolution from record albums to CDs to MP3 files for music has changed the way that people buy and listen to music, i.e., by song chunks rather than collection chunks. Mashup and reuse alternatives raise new levels of consumer participation with the artifact. I claim that electronic information artifacts lead to more active consumption and reuse.

4.3.4 SHARE

One of the hallmarks of human progress is the transfer of knowledge across generations. Information artifacts are the mechanism of this transfer. There is substantial human effort given to creating information artifacts. Even in cases where information is not created to share with future generations, archivists, museum curators, or archaeologists collect such artifacts to document an individual's life and role in history. Whether intended or not, the sharing phase of the information life spiral is important because it has spawned powerful industries such as publishing, broadcast media, and paper or digital postal services, and sustains a range of cultural memory institutions such as libraries, museums, and archives.

The distribution of information artifacts requires three kinds of activities: copying, delivering, and record keeping. Paper-based artifacts must be printed, which is typically done by specialized entities that take the master artifact and transform it into type that is used to press ink on paper. In traditional publishing, metal character forms were laid out on printing presses, ink was applied and paper run through the press and then cut and bound. Today, computer files of the master artifact are used to control printing presses that spray inks onto paper. These physical processes are repeated to make a 'run' of the artifact. A scholarly book may have small print runs (e.g., 500-2000), whereas popular novels will have hundreds of thousands of copies in a run. An important property of electronic artifacts is that exact copies can be made automatically at minimal cost. Thus, the copy activity of sharing becomes trivial. This property strongly affects the meaning and motivation for copyright and shakes the foundation of the publishing and broadcast industries.

Delivering information artifacts requires getting copies of the artifact to individuals. Traditionally, this requires warehouses for storing the artifacts, and transportation systems to move them to different localities. Just as with the copy activity, thousands of jobs and large investments have been required. Electronic networks and the WWW change this delivery activity. An individual can deliver an electronic file artifact to thousands of people on an electronic list with a few keystrokes, or post the file to a website, blog, or social media site thus giving access to potentially millions of people.

Keeping track of where artifacts go is the third sharing function. Keeping records of who should get the artifacts and the payments received is crucial to sustaining the sharing enterprise economically and operating within the many laws of modern society. For-profit publication industries must have billing records to operate and even non-profit organizations must at least keep records of production and delivery to meet regulatory standards.

Just as electronic artifacts and electronic networks change the fundamentals of copying and delivering artifacts, the record-keeping functions are also evolving. At first look, it would appear that an electronic artifact posted to a blog that is open to the public requires no record keeping. A more reflective look, however, brings into view the many kinds of logs that computer systems generate for a variety of purposes (e.g., backup, recovery, advertising, and data mining). More importantly, the nature of electronic networks and the artifacts they carry introduce new levels of interactivity where people who consume artifacts can easily and immediately provide feedback on the artifact they consume. Comments, ratings, and such derivatives quickly augment the artifacts and make the sharing process more dynamic. Today, these interactions are used by creators but also by different layers of services in the WWW to mine behavior patterns for a variety of purposes (e.g., to generate ads or to collect research data). It is inevitable that the economic, legal, and political implications of these changes in the sharing phase of the information life spiral will lead to new record keeping requirements. For example, journalists increasingly mine public information artifacts and file requests for access to government or corporate records under Freedom of Information rules or commercial regulation authorities. This leads to new layers of services to manage and monitor the flow of information artifacts.

4.3.5 PUBLIC MANAGE

Managing collections of information artifacts over the ages has given rise to libraries, archives, and record centers. It has also given rise to fields of study such as information and library science that develop techniques and systems for managing huge volumes of information artifacts. Key activities associated with managing large collections of information artifacts include: collection development, indexing, retrieval, and preservation. Librarians, archivists, and records managers make decisions about what artifacts to collect by developing and following collection development policies. These policies may be driven by community and popular culture interests, business objectives (e.g., customer records, product specifications), legal mandates (e.g., government retention schedules, corporate financial records), or scientific and scholarly progress. Clearly, these policies also reflect back on the creation process as they serve to stimulate new artifacts over time.

In the case of electronic artifacts, data warehouses, digital libraries, and institutional repositories have arisen to manage collections. Although the ingest processes may change from trucks, loading docks, and shelves to electronic transfers over wire or fiber, collections are built according to collection policies that are rooted in organizational or community interests and values.

As collections grow, it is essential that indexes are created to help people locate individual artifacts. Indexes are built on common artifact feature sets (e.g., words, names, dates). These feature sets are called metadata elements and may be extracted from the artifact (content-based) or added by people (e.g., topic, genre) or machines (e.g., time stamp, aperture setting) either at creation time or afterwards. Over time, libraries created metadata standards to label individual artifacts and ordered lists of metadata tokens that link to physical locations. To help scale to millions of books, compressed codes such as Library of Congress (LC) subject headings represent the major topic contained in the artifact and order of the artifacts in the collection. For example, a particular book on information retrieval might be assigned a LC call number such as TK5105.884.C765 2010, which assigns it to a shelf position with books on the same general topic. Each library uses a particular classification system (in the United States, Library of Congress and Dewey classification systems are most commonly used) so the call number is specific to a collection. However, modern libraries share cataloging information and call numbers, so libraries that use the Library of Congress system will use the same call number for this book. In addition to a call number that is keyed to a specific library classification system, books also have identifiers assigned by their publishers (ISBN number for this same book is 0136072240, which codes for publisher as well as item in the publisher's catalog) and by shared cataloging services such as OCLC (this book has OCLC number 268788295). A search in the online catalog (the electronic version of the card catalog) for the term 'information retrieval' would yield a set of books and their call numbers that in effect point to specific shelf locations via the call number.

There should be many ways to find a single artifact so libraries create indexes for authors, titles, and topic. If the book text is in electronic form, every word can be indexed to support full-text searching, which provides a plethora of pointers to a shelf (or disk track) thus increasing the chance of finding the artifact with many different searches. Of course, full text indexing on all words in all books is not effectively selective (returns too many false positives for keyword search) and so more selective indexing is done in practice. For images, music, or video artifacts, other kinds of features such as color, texture, or wave form are used for indexing (see Christel, M. [2009] and Ruger, S. [2009] for lectures in this series related to video and multimedia retrieval). The success of search engines such as Google, Yahoo, and Bing are due to indexes for web artifacts that leverage words as well as the hyperlinks among those artifacts.

Indexing is the basis for retrieval. From the perspective of people who search for information artifacts, the challenge is to articulate a need and receive the most appropriate artifact in return. Thus, managers of collections must provide search systems that help people map their information needs to artifacts via the indexes for collection. Libraries provide human reference services to elicit needs (see Lankes, D. [2009] in this series for a lecture on electronic reference services) as well as catalogs (most paper catalogs have been converted to online public access catalogs—OPACs) to help people retrieve relevant artifacts. WWW-based collections offer retrieval via web browser-based interfaces that support query-based or facet-based retrieval (see Tunkelang's lecture on faceted search in this series).

Storing and preserving information artifacts requires space with proper controls for security and environment. Protecting artifacts from theft and damage causes some managers to restrict access (e.g., in some libraries shelves are not open access) and embed sensors in the artifacts and entrances. Different substrates require specific kinds of environments. Paper is flammable, susceptible to mold, and deteriorates over time. Thus, temperature and humidity controls are important.

Electronic artifacts such as computer files present new challenges for preservation because formats and technologies change rapidly. The archives subfield of digital curation addresses problems such as digital forensics (recovering files that have damaged bits), scalable storage (scientific data files in the petabyte range exceed parameters of typical operating system and software parameters), and authority control (regularly checking the veracity of the bits). Managers are wise to 'weed' their collections by deleting artifacts that are no longer needed. Retention policies guide 'weeding' to save people time and effort at retrieval (reduce irrelevant or redundant items) as well as to save storage and index processing[21].

4.3.6 PUBLIC USE

People use shared information artifacts for many reasons such as planning, verification, decision making, and learning. The changes that electronic information artifacts bring to personal use obtain for public use as well but are extended in two important ways. One extension is use of the relationships between artifacts and the usage patterns that accrue for these artifacts. The relationships expressed in hyperlinks, feature co-occurrence (e.g., same words, color), or annotations by others are used for retrieval (e.g., page rank) but they are also used by people who follow these relationships while they browse collections of artifacts[22]. The patterns of access, comments or tags are used to discover related artifacts (e.g., through recommendations). This use of artifacts serves not only the user but also others and adds context to the artifact that in essence changes it through use over time.

Secondly, the development of social media methods and tools in the WWW substrate opens new scales of public sharing and blurs many of the traditional spatial boundaries between private and personal information artifacts. These social media substrates are highly malleable for information creation, effortless for sharing, and increasingly persistent and unmanageable. They provide substrates for augmenting and creating personal identity. With the fundamental properties of electronic information artifacts, they give rise to a new form of information (Voice5), proflection of self that is treated in Chapter 6.

4.4 ARTIFACT EVOLUTION

Over time, everything changes. Marble artifacts eventually will crumble and books wear out with use or environmental conditions. Changes in artifacts have tended to be slow with respect to human life spans. Traditionally, an information creator who intends an artifact to persist can use a substrate and methods that insure that under ordinary use conditions the artifact lasts much longer than a single life span.

Artifacts do not stand alone. To be useful they must have metadata that makes them findable. They must also exist in contexts that make them useful. Local context associated with the creator and the artifact life spiral helps users to understand intention and make interpretation. Global context situates the artifact life spiral in time and culture. For example, an epitaph carved on an ancient gravestone may maintain the characters chiseled by the creator two thousand years ago, but the local

[21] See Rajasekar et al. [2010], in this series for a technical approach to specifying various digital curation policies as computable rules.

[22] See White and Roth [2009], in this series for a treatment of exploratory search.

context associated with grieving parents of a son fallen in battle and the global context of the battle's significance in history must be captured in related artifacts for modern people to understand the intention of the artifact creator.

Figure 4.3. illustrates how artifacts are embedded in context and how they evolve over time. The artifact and the metadata associated with it are separated by a dotted line to show that the metadata can be embedded in the artifact (e.g., the title of a book is part of the book) or it can be distinct (e.g., a bibliographic record for the book can exist in a physical or electronic card catalog). The artifact and metadata are situated in local context that may overlap with the local context. For example, the author's name provides local context, serves as metadata, and also is part of the book. The educational and employment history of the author may strongly influence the creation of the book but not appear in the metadata or in the book (beyond what might be in the brief biography on a book jacket). All these elements are embedded in global context of the creation time, some of

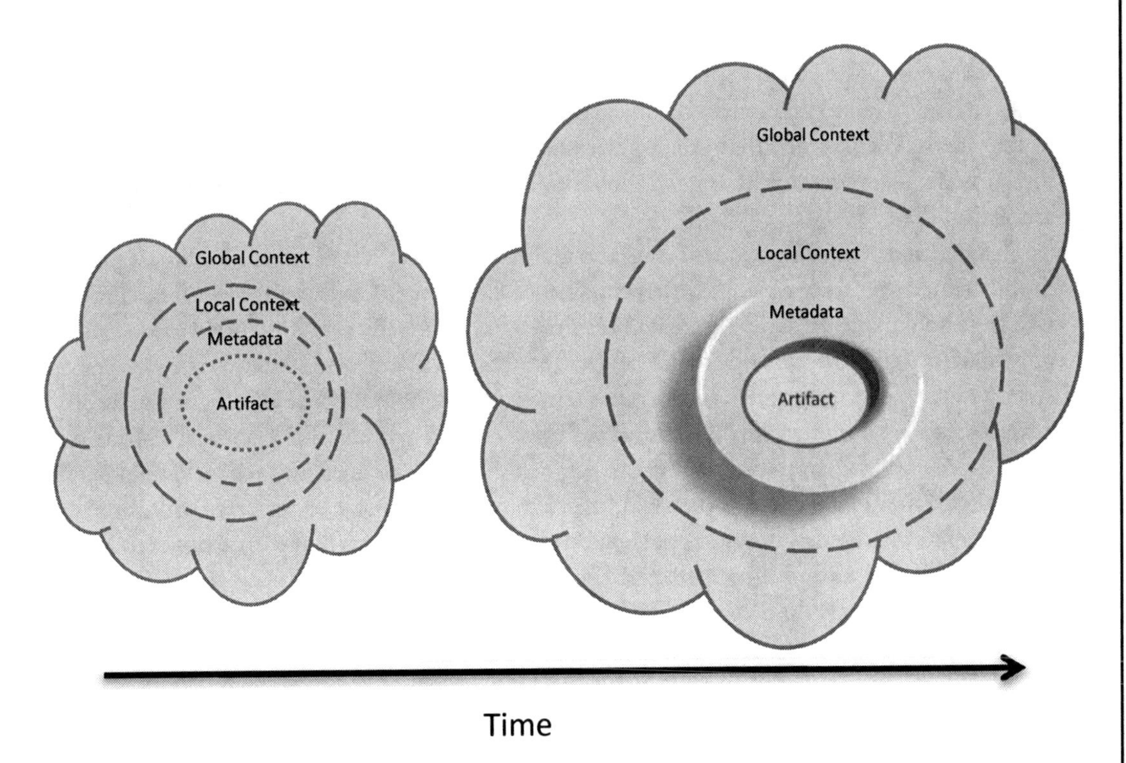

Figure 4.3: Information Artifacts Evolve over Time.

which may be represented in the artifact but typically is not. The right side of Figure 4.3 shows that over time the artifact evolves. The artifact acquires a history. The artifact itself may have annotations or show usage. More metadata may be extracted or added. A significant amount of local context may accrue as the artifact is used, reviewed, discussed, and derivatives are created. The global context continues to expand with the march of human history. It is apparent that electronic artifacts have the potential to more easily blur the boundaries of artifact, metadata, and local context and for the metadata and local context to accrue much more rapidly. The possibility of collaborative creation where iterations occur in short durations becomes possible (e.g., Wikipedia, collaborative projects via shared tools).

This evolution was made explicit as part of a digital preservation project. We harvested YouTube videos on a set of specific topics on a daily basis over an 18 month period (Marchionini et al. [2009]). The artifact (video file) and metadata (description, category, location, etc.) were provided by the video creators. Viewing patterns (number of views over time), rank in the results list each day, comments, ratings, and mashups created as derivatives were examined as indicators of impact. For the 2008 U.S. presidential election, viewing patterns for popular candidates over time reflected both voter interest and the sophistication of the campaigns to use the WWW to garner support for candidates. The patterns and comments provide local context that was captured for future generations to examine. The global context of the election (e.g., what the popular media stories were at the time) could also have been systematically gathered, as we correlated blog postings on the candidates to the videos. We assumed that the popular media (newspapers, TV news) will be archived so that in 100 years historians will have good global context for situating the local context, metadata, and artifacts of 2008.

One kind of information artifact that is mainly enabled by electronic technology is the personal profile created over time as people interact with various systems. Consider the carefully curated set of web page bookmarks that people create over time as a continually evolving information artifact. We only tend to recognize the importance of these evolving artifacts when we upgrade systems or have system failures and either have to migrate them or we lose them. More subtle are the information artifacts that are automatically created as we interact with systems. For example, a spam filter for email 'learns' what characteristics of email messages we wish to keep and which should be filtered. Each time a new email client is installed, the training process must be restarted unless the existing profile can be migrated. Beyond our personal systems, many systems that we use remotely build some kind of profile for us and these personal histories may survive long after we do or disappear after each session. Not only do these profiles have highly variable half lives, but we often are not aware that they exist. There is an interesting connection between our memories, which are internal and dependent on our life energy and our histories which are external and independent of our existence once they are created. In fact, social media services such as Facebook have difficulty dealing with increasing numbers of accounts that are held by people who have died (e.g., should they be deleted and what about all the postings from and to others?)[23]. The evolution of our personal profiles and their significance to our lives define a new kind of information that is treated in Chapter 6.

[23] Kelly, M. [2009] presents the Facebook case for memorials that persist after people die.

4.5 ELECTRONIC INFORMATION ARTIFACTS AND HUMAN EXPERIENCE

The clear distinctions between the thoughts that occur as one reads a book (information in the head), the physical book itself (information as artifact), and the act of writing the words that go into the book (information as communication process) begin to blur when the book is an electronic document (e.g., blog, social media posting) that allows annotation by the reader, the author, and others and is open to electronic agents that index, relate, and link the document. Electronic tools allow people to quickly and easily cycle through the information spiral to revise and augment information artifacts. For example, it is easy to capture (create) and email (a communication act) a photograph (an artifact) to someone. Thus, the execution of the phases of the life spiral is compressed and, in turn, changes the user experience to be faster and more expansive for collaboration and sharing.

What is less clear is how these tools influence the mental representations the emailed photograph causes in the mind of the viewer. Receiving a photo of my granddaughter, I can display it on a large or small screen, alone or in montage with other photos, and can crop or manipulate pixels in a variety of ways to present it persistently or dynamically under conscious actions or automatic rules. It is highly debatable whether a word processor makes one a better writer, a digital camera makes one a better photographer, or a spreadsheet program makes one a better accountant, or whether these tools make us better consumers of information. It seems certain that electronic tools amplify the rate of information interaction. However, it remains to be determined whether these tools truly augment our intellects as argued by Bush, V. [1945], Engelbart, D. [1963], and others in the past half century. With electronic tools we can be as actively engaged in controlling representation as we are in interpretation and meaning making.

To understand the implications for human experience, we summarize the features of electronic information artifacts.

- Require power. This makes them ephemeral and necessitates secondary artifacts for storage.

- Substrates are intangible to human perception. The substrate changes are imperceptible to human senses; the codings are orders of magnitude smaller[24] and in spectra beyond the human sensory system. This necessitates rendering devices.

- It is easy and cheap to make perfect copies at scale. This disrupts traditional publishing models of production and copyright law.

- It is easy and cheap to massively transfer copies at scale. This facilitates new kinds of sharing and collaboration. It also disrupts the publishing industry.

[24] Magnetic, optical, and electronic artifacts have scales in the millimeter (10^{-3}) to nanometer (10^{-12}) range, dramatically smaller than human perceptual scales (roughly 10^{-3} to 10^{3} meters).

- Require small amounts of physical storage space. This mitigates the need for large physical warehouses[25] but leads to new challenges for management and preservation.
- Support multimedia. This adds new richness to the information experience but also add more kinds and layers of coding and more rendering components and processes.
- Incorporate behavior (are interactive). This may be the most significant characteristic because it changes the way that people experience the artifact. Interactive artifacts also disrupt traditional industries (e.g., consulting, product sales) and laws (e.g., state taxes on national or international sales) that have evolved over time.

These characteristics of electronic artifacts have implications for human information experience. Some of the implications follow logically by extrapolation and some are emergent and transformative. The disruptions to the information industry that size, reproduction, and instantaneous global transfer bring are not surprising, although the transformations will take place over decades.

The WWW has challenged services such as travel, real estate agencies, and mass media (e.g., newspapers, TV) to adapt their methods of operation. Electronic artifacts have created unprecedented opportunities for planned and emergent collaboration. Collaborative work tools allow document sharing so that people at remote locations can participate in shared work or play. Services such as Flickr and Wikipedia allow people to explicitly share photos and create and sustain encyclopedic entries for concepts respectively. New kinds of collaboration are also made possible. For example, people are now able to collaborate on information search tasks ranging from planning vacations or major purchases to health care. Research and development on collaborative information retrieval (see Morris and Teevan [2009] in this lecture series) has yielded new kinds of services to explicitly support this kind of collaboration.

Emergent collaborations arise as a result of the scale of personal collaborations in an open environment. The open source software movement is perhaps the best example of massive collaborations in an open electronic environment. See Raymond, E. [1999] for an insider's view of the development of Linux and Deek and McHugh [2008] for a human-centered study of many open source technologies. The willingness of scientists to share large-scale scientific data sets and increasing pressure from funding agencies to share these data have spurred scientific discovery. The human genome project is perhaps the best known international sharing effort; however, examples abound in every scientific discipline. The edited volume by Hey et al. [2009] describes scientists working with computational scientists to integrate, analyze, and visualize large scale data in a variety of areas. Shneiderman, B. [2008] has argued that even greater discoveries will come from the social data accruing in the WWW as social scientists learn more about human social behavior by studying these data sets.

Even K-16 education has begun to leverage collaborative learning. Traditional science education has used empirical investigations in which students re-create a discovery made long ago.

[25] However, as electronic artifacts scale, expensive data centers are required. For example, today's search engine companies invest half a billion dollars each for data centers and build several of these each year (Dan Reed, National Academies of Sciences Board on Research Data and Information meeting oral statement, June 4, 2010).

Simulations allow students to (hopefully) reach the same conclusions more effortlessly and efficiently. What is gained and lost is hotly debated in the learning sciences. WWW-scale collaboration offers the opportunity for students and citizens to participate in mass-scale data collection (e.g., Sloan Digital Sky Survey; the Cornell Lab of Ornithology operates several bird watching surveys), and to use existing data sets to make new discoveries and investigate social data sets of personal interest. Such applications also raise new kinds of questions. Just as the use of simulations rather than first-hand experiments generate controversy, students using large scientific data sets, and especially social media data sets also is controversial. Issues arise such as whether students learn to think and discover for themselves or to simply retrieve what is known. Also behaviors such as cyberbullying evolve because social media do not distinguish the implications of good or bad human behavior, and thus require new kinds of self-discipline.

Examples in other domains include leveraging social data such as photo tags to enhance search results and constructing 3D models of popular historical sites by integrating thousands of photos contributed by different visitors (e.g., Agarwal et al. [2010]). The possibilities for new advances as well as for abuse of emergent collaboration capabilities are substantial, and it will likely take generations to fully understand the impact.

It is clear that electronic technologies have greatly expanded the range of information as artifact. Electronic information artifacts have become services (e.g., search engines). These new artifacts have, in turn, had large impacts on how information is used (e.g., ubiquitously rather than in specialized locations) and in the way it is stored, replicated, and distributed, managed, and searched.

CHAPTER 5

Information as Energy

Information is a kind of energy that drives learning, comprises plans, and effects changes in physical or conceptual states. Energy is what effects changes in nature at all levels, from the subatomic to cosmic. There are many kinds of energy (e.g., mechanical, electromagnetic, chemical, nuclear, thermal), each with basic properties and measures used to define and study it. Energy can be active (kinetic) or stored (potential), transformed, and measured. We understand energy through the quality and quantity of change it effects. For example, a basic measure of energy is the joule, which combines mass, distance, and time qualities and quantities (a joule is a one newton force that moves an object one meter; in effect, the amount of force required to accelerate one kilogram one meter in one second on earth.)

For informational energy, the qualities and quantities are less well-defined. If we treat reduction in uncertainty as a state change (the work done by informational energy), then probabilities of change can be used as measures for information. For mental or social states, the state change qualities are much more subjective and not (yet) reducible to probability values. One scientific view is that we may be able to determine such values for mental state changes through techniques such as functional magnetic resonance (e.g., Mitchell et al. [2008] mapped word recognition activations in the brain to stochastic functions used in machine learning). Others believe that brain activity is not sufficient to explain mental state change. Likewise, defining and measuring state change in social states (e.g., human recorded knowledge) presents both qualitative and quantitative challenges. In this chapter, we first discuss changes in physical states from the technical perspective of information theory. We then turn to changes in mental and social states.

5.1 CHANGE IN PHYSICAL STATE: REDUCTION IN UNCERTAINTY

Although this lecture focuses on information senses from a human-centered perspective, it is important to consider generalized senses that cut across different human senses and that support system design. The most important technical sense of information of the 20th century was the notion that information is defined by a reduction in uncertainty. Shannon, C. [1948] laid the foundations for a theory of information based on his desire to make telephone signals effective and efficient. He proposed that information can be defined as a change in state. He suggested that the key elements in a communication system are the source, channel, and receiver. Processes of modulation occur from source to channel, and channel to receiver, and are influenced by the presence of various kinds of noise in the channel. The source consists of a finite set of information units (well-defined alternatives), which in the simplest case are binary values (bits). The source codes these units using a protocol appropriate to the channel (e.g., high or low voltages in the case of analog signals or bit packets in the case of digital signals sent over the Internet) and the channel propagates the encoded signals to a receiver that decodes them. The receiver is in some state with respect to the well-defined alternatives before the communication begins, and the newly arriving units 'inform' (change the state of) the receiver by decreasing the uncertainty about the alternatives in the receiver.

Suppose you have invited several families to a dinner party and await messages (information) from the invitees. Consider one source (family) that has a set of four alternatives, representing whether 0, 1, 2, or 3 people in the family will come to dinner. First assume that any of the alternatives is equally likely. Before you receive the reply for this invitation, you have maximum uncertainty about how many people from the family will come. You have a one in four chance of guessing how many people will come from this family. When the reply comes (the source of the information codes the message and sends it via the channel and you receive it), you are now 'informed' about this event. You have reduced your uncertainty from a best case of .25 ability to guess a correct event to knowing exactly (1.0 ability to tell) how many people from this family will come.

Clearly, this simple example ignores the kinds of inferential thinking that a human would do to treat the probabilities as not equally likely. For example, you might know from experience that this particular couple seldom brings their child but often comes to parties as a couple and, occasionally, one will come alone. Suppose you estimate the respective probabilities[26] as $p(E_0) = 1/8$, $p(E_1) = 1/4$, $p(E_2) = 1/2$, and $p(E_3) = 1/8$. Now, when the reply arrives saying that all three family members are coming, you are more surprised than if the reply says that two are coming—you thus have a corresponding larger reduction in uncertainty. For this example, the quantity of information for the single case that all three are coming should be more than the quantity of information for the more expected case that the couple are coming without their child. By considering this quantity to be reduction in uncertainty, we can see that the more surprising event carries more information. When the probabilities are not the same for all events, we need a better estimate for the initial amount of uncertainty and for this, Shannon identified the concept of entropy. Before discussing entropy, we formalize the basic concept as follows.

A source has a number of items, each with associated probability estimates for selection. The receiver has uncertainty about which is selected. $P(E)$ is the probability that event E will be selected from the possible alternatives in the source. The logarithm (log) of a number for a given base is the power to which that base must be raised to yield that number. Thus, the log10 of 100 is 2 because $10^2 = 100$. Because it is convenient to code telephone signals or other signals onto only two values, we typically use the base of 2 because it maps directly to the two (binary) states of digital representations. Using base 2, counting proceeds as follows, 0, 1, 10, 11, 100, 101, 110, 111, 1000, etc. and $2^0 = 1$, $2^1 = 10$, $2^3 = 100$, $2^4 = 1000$, etc. Each of the columns (places) represent one binary digit (1 or 0) ('bit' is a shorthand for binary digit), which serves as the unit of information.

[26] $P(E_0)$ here means the probability of the event that 0 people come to the dinner.

Using these concepts of probability and logarithms, Shannon defined the information content in a single event as $I(E) = -\log_2(P(E))$. For the party invitation example with 1/8 as the probability that all three family members will come, we have $I(E_3) = -\log 2(1/8) = 3$ (because $8 = 2^3$). Note that the negative sign in the general formula is due to the nature of logarithms, which have negative values for fractions (e.g., log(2) of 8 is 3 but log(2) of 1/8 is -3). For the case of only the couple coming, we have $I(E_2) = \log 2(1/2) = 1$—less surprise that the couple comes without their child.

Thus, we have a base for measuring the quantity of information in a single, simple, and discrete message. Shannon aimed to characterize systems that serve more general cases and also to treat transmission rates, channel capacity, and coding to improve efficiency and accuracy. To build on the basic formulation for a single event, we return to Shannon's introduction of entropy to describe the uncertainty in a given set of choices rather than a single event. He defined entropy as $H = -\Sigma p(e) \log(p(e))$. In our party invitation example with probabilities 1/8, 1/4, 1/4, and 1/8, we compute the total amount of uncertainty (entropy) as $-(1/8^* - 3 + 1/4^* - 2 + 1/2^* - 1 + 1/8^* - 3) = 1.75$. When we compare this to the entropy in the situation where there are equal probabilities for all four responses (entropy=2), we see that there is more uncertainty in the overall situation with equal probabilities than when some additional distinguishing probabilities are available. This agrees with our intuition and Shannon showed that we have maximum entropy when all events are equally probable, and decreasing amounts of entropy as the probabilities are more distinctive. The more we know about the relationships (probability distribution) in the source, the less surprised we will be when information arrives (overall). Thus, the mathematics of Shannon agrees with our intuitive notions that the more we know about something, the more predictable it is to us.

This simple example demonstrates the basic concepts of Shannon's information theory. Shannon used it to describe measures for conditional events, streams of messages, and situations that involve infinite sets of alternatives. His work has been used as the basis for modern telecommunications systems ranging from telephones to the WWW and is used in biology and other fields to give computable predictions for system dynamics.

The examples above apply to what Shannon called information in a discrete, noiseless channel. He extended the work to noisy channels, which led to important breakthroughs in coding theory and message estimation. Extending the work to the case of continuous channels provides the basis for biological treatments of information flow in organisms.

See Shannon's original paper [Shannon, C., 1948] and the book with Warren Weaver [Shannon and Weaver, 1949] for the original full development of these ideas. Hundreds of books and papers have explained and extended this perspective on information. See Applebaum, D. [2008] for a readable introduction to modern probability and information theory. Roederer, J. [2005] offers a treatment from the viewpoints of physics and biology.

Shannon used this notion of reducing uncertainty to develop mathematical models of information flow that allow engineers to predict how much information a channel can transmit. He also did groundbreaking work on theories of coding information for channels. Shannon and Weaver acknowledge that this model of communication addresses the engineering aspects of communication but not the semantic aspects of communication that people are concerned with in their usual communication events. It does, however, serve two important purposes: First, it provides a precise model for building communication systems. Second, it provides a simplified model for thinking about information as the act of informing others (reducing uncertainty).

5.2 CHANGE IN MENTAL AND SOCIAL STATES

As discussed in Chapters 2 and 3, people learn by recalling or perceiving information. The changes from one state of thought to another require some kind of energy form, regardless of the philosophical perspective taken. Biologically, neural pathways depend on biochemical energy flows across neural cells and neuroscientists are making some progress in defining and measuring these energy flows at the cellular (there are entire journals devoted to cell signaling) and organism levels. For example, Laughlin et al. [1998] measure adenosine triphosphate (ATP) consumption as the unit of energy (information) transfer across cells. Their information energy model predicts that 10,000 ATP molecules are required to transfer one bit at a synapse, and that overall photosensor rates of 1000 bits per second are typical in neural signaling.

Psychologically, change in mental state is described as different kind of actions: 'firing' of neural networks; execution of production rules; mental 'flows,' or 'runs' of mental models. Regardless of the model of cognition, some kind of action (energy) is required to effect these changes, and we can consider these as information energy. At present, we describe these energies as learning and measure effects with cognitive tests that aim to determine the effect of change in mental state.

There is considerable work in the educational field to define and measure cognitive load, which might stand as a more direct measure of information energy within the learner. Cognitive load is the burden that one's cognitive system bears while engaged in a task (Paas et al. [2003]). The more difficult a particular task is, the more the task taxes one's mental resources, consequently leaving fewer mental resources available for other functions. If we replace 'difficult' with how much reduction in uncertainty (information) a task entails, then measures of cognitive load could be used as measures of information.

Sweller, J. [1988] defined different types of cognitive load that influence learning. Intrinsic load refers to the effort needed to understand the content (information), which varies according to the level of abstraction, granularity, and general complexity. Extraneous cognitive load refers to the load that the medium or delivery system puts on the receiver. Instructional designers aim to create systems that minimize extraneous load while causing sufficient intrinsic load to maintain attention and thus change mental state without overloading or underloading the learner. Achieving this balance requires knowledge of the learner's existing state of knowledge and thus drives strategies for individualization. Researchers measure cognitive load with indirect measures such as questionnaires, and with direct measures such as giving secondary tasks (e.g., count by twos while searching for visual targets), and measuring time differences in the primary task. More recently biometric or fMRI assessments are used to assess cognitive load. Most of this work has been done in the context of education and learning; however, it is reasonable to expect that such approaches can be useful for measurin g information energy effects in humans.

Rather than individual mental states, we can consider the state of a social system (group, organization, humankind) and what causes changes in these states. Distributed cognition (Hutchins, E. [1995]) posits that human knowledge is held in the aggregate of multiple minds and the artifacts of mankind. Clearly, this is constantly changing and the changes must be the result of information energies. We might limit examination of these states to external global artifacts (e.g., all the physical and digital libraries, all the news broadcasts around the world) and describe the qualities and quantities related to their evolution. Wikipedia, for example, is constantly in flux (changing state) and represents a large social effort that results in an evolving information organism.

We may limit the social state to the aggregation of all the mental states of people–what everyone is thinking at a given point of time. On ordinary days, the collective state may be near random (maximum entropy), but when a global event (e.g., massive earthquake, World Cup Soccer championship) is happening, then the collective state may become more unified and therefore have less information potential. Experiments in crowd sourcing are beginning to yield examples that may be useful to understand these changes in collective consciousness. As considered in the next chapter, we might extend both of these to include what is added by computational agents and the linkages within the WWW. In all of these cases, information flows are continual, ambient, and crucial to change.

CHAPTER 6

Information as Identity in Cyberspace: The Fifth Voice

In the previous chapters the different classical senses of information were discussed with an eye toward new kinds of information artifacts, information interactions, and human information activities that emerge with electronic technology. In this chapter I argue that the evolution of electronic systems has yielded a new kind of information artifact substrate instantiated in the WWW, and that human adaptations to their ubiquity is defining a new information space for human interactions. This new environment is termed cyberspace and exists between our physical and mental spaces. Cyberspace is populated by people, electronic information artifacts, computational agents (programs), and traces of human activity. Cyberspace has become an instance of collective human knowledge. It is dynamic and more expansive than any single mind or institution can manage. Partitions of cyberspace at any instant in time are new kinds of information artifact. The partition of cyberspace that pertains to an individual represents that person's identity in cyberspace. Human interactions with others and with computational resources in cyberspace determine alternative expressions of personal identity that persist, morph, and propagate as a new kind of information that I call *proflection of self.*

Proflections represent our personal identities in cyberspace and emerge as the products of our conscious and unconscious actions in cyberspace. They are a product of the myriad collaborations and interactions we have with people and computational agents. These interactions may be intentional or not, and they coalesce into dynamic personal profiles that affect how other people and agents understand us and influence our subsequent activities in both cyberspace and physical space. Increasingly, the boundaries between cyberspace and physical space are blurring, which makes our identities in cyberspace especially important.

This chapter first presents an overview of the notion of cyberspace, then elaborates the nature of information interaction, and the importance of personal identity. The concept of proflection is then presented as the combination of different types of projections and reflections. The chapter concludes with implications for learning, work, and leisure.

6.1 CYBERSPACE

Cyberspace is a general term that is used to encapsulate the confluence of people, global electronic networks, information artifacts generated by people and machines, and software agents that act in these networks at any moment. People act in cyberspace through technical interfaces that support input (e.g., keyboard, game controller, sensor) and output (e.g., LCD display, audio speaker). The networks include the following: telecommunications channels that guide the flow of energy; layers of hardware such as servers, storage arrays, and routers that act as physical nodes; various protocols that insure exchange; and suites of software that control the hardware and instantiate the protocols for energy (information) flow. The information artifacts include the entire range of databases, webpages, videos, binaries (e.g., programs, apps), directly addressed messages (e.g., email), and public messages (e.g., blog posts, tweets, wall postings) that people purposefully inject into the network, as well as a variety of logs and computations that machines add. Software agents include software that index, sort, search, pattern match, aggregate, compare, and filter the information artifacts and activity streams of people and other software. These interacting components cause cyberspace to be in constant flux.

Cyberspace is information energy that represents a new kind of global knowledge. Just as no single mind or library can store all of human knowledge, no single entity known today can perceive or store all of cyberspace at any time interval. Furthermore, cyberspace is dynamic because large portions of humanity and our computational agents continuously act within it. A search engine interacts with human searchers to represent a subset of cyberspace that conforms to a well-specified constraint (query); however, the same query moments later (or executed by another person in a distinct context) may yield a different subset. For example, a query about a natural disaster (e.g., earthquake) or a famous person will return very different results before and after an actual event such as a disaster or the passing of the person. The scale, dynamism, and diversity of potential and active energy (information artifacts) and number of people and software agents make cyberspace a state of the world that is beyond any single human or machine. Nonetheless, the unprecedented capabilities we have to create selective subsets gives us extraordinary power. These instantaneous subsets of cyberspace are a new kind of information artifact.

Humans are embodied in physical space but also exist in personal and public mental spaces. As we read, view a film, think about the past or the future, or imagine worlds afar, our lives during those moments exist in those mental spaces. For example, Sturm, B. [2000] calls the mental state that engaged story listeners enter a 'trance'—an altered state of consciousness. Sane adults move effortlessly between physical and mental spaces without confusion. Young children and some psychologically disturbed adults confuse these spaces. The totality of mental spaces captured in information artifacts is the totality of human recorded knowledge. This is not the totality of collective mental space. Figure 6.1 places cyberspace between physical and mental space on one dimension, and between personal and public space on a second dimension.

The figure shows the personal pole of the personal-public dimension in gray because it is difficult or impossible to maintain information privately once it is placed in cyberspace. In the physical world, we control analog objects (e.g., an automobile to travel, a diary to record our experience, a cell phone to communicate, our bodies to sustain life). We are able to recuse these objects from public places and verify that they are indeed solely where we put them. In cyberspace, we control bits (actually higher level abstractions of bits) to accomplish a variety of goals (e.g., documents to learn and to communicate, code to execute well-defined tasks, video games to entertain ourselves). Increasingly, the objects in these spaces share digital components as the number of computational

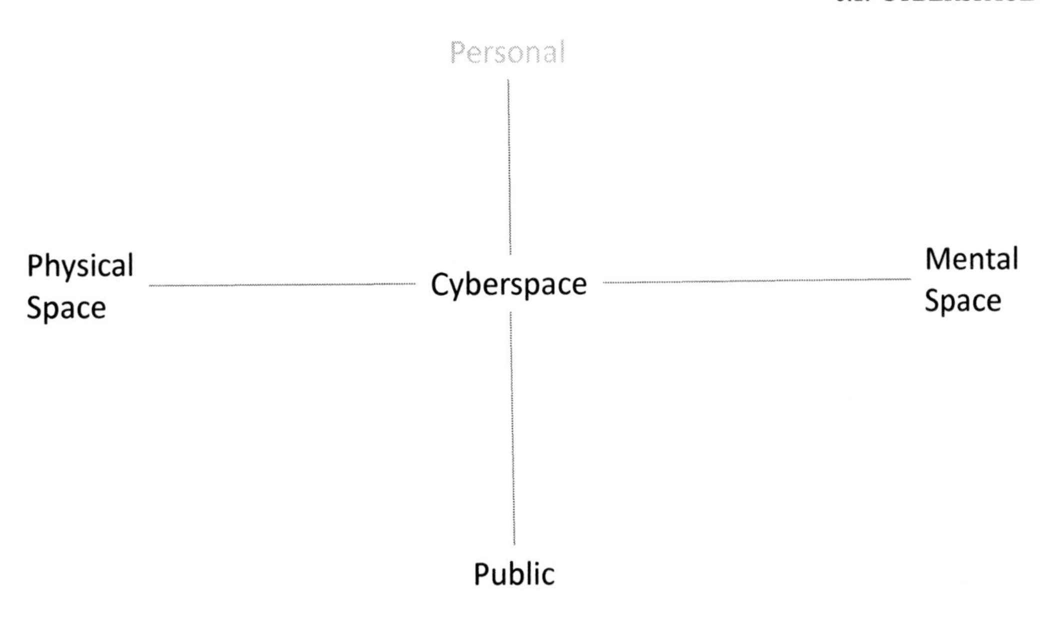

Figure 6.1: Cyberspace Conceptual Dimensions.

units in our lives grow and communicate with each other and as we take sensors into our cars, homes, and bodies. These components are connected in myriad ways across data grids in combinations that we cannot trace. Thus, we may save an electronic file in a private region of cyberspace, but backups and archives must be created, and we only have trust in the datagrid provider that these bits are secure and not accessible to others.

We also live in private (e.g., our homes, autos, and minds) and public (e.g., offices, malls, buses, and online forums) spaces[27]. We have private accounts, public blogs, and various levels of open and secure wireless and wired networks in our offices and homes. We have public and private collections of information artifacts, and although we might expect our actions in each to inherit the same public/private properties, the fluidity of cyberspace does not require this. In fact, once in cyberspace, our artifacts propagate and even our actions (e.g., searches for a pair of shoes at an e-commerce site) that are semi-private between us and the merchant are used in collaborative filtering or data mining applications. These applications depend on our semi-private (e.g., commercial) actions to make recommendations to others with similar profiles.

6.1.1 CYBERINFRASTRUCTURE INTERFACES

Civilizations have created many infrastructural components to make physical space more convenient. Water, power, and transportation grids make modern life styles possible. Cyberspace also requires

[27]The public/private dimension is more a continuum than a bipolar set of states.

underlying components that support convenient human activities. The hardware, software, protocols, data sets, and human workers who make cyberspace possible are termed cyberinfrastructure[28]. Cyberinfrastructure is developing rapidly but many challenges remain, especially with respect to the implications of how it affects our lives. In the following section I highlight one area of particular human-centered interest: user interfaces for working in cyberspace.

In spite of excellent developments in cross-platform languages, we seem destined to have as many ways to interact with electronic information as we do with the physical world. Cyberinfrastruture must support this diversity. To illustrate this, consider a small set of interface needs that are desirable in a permeable space, defined by private-public and physical-cyber-mental dimensions.

We need interfaces that help us coordinate and manage our many devices and services. When we move from scientific data sets in a digital library to streams of new data from instruments or from `Amazon.com` to our files in our private data grids or across different email and messaging clients on our phones, computers, and intelligent buildings, our preference settings, collaborative filtering profiles, histories and bookmarks should move with us. We might imagine common interfaces that we adopt and that move with us to various applications. Today, I can move my information artifacts (files) from mobile to laptop to desktop devices regardless of the operating systems but must use distinct user interfaces for each system. A mouse move on one equates to a gesture or utterance on another. We need new sets of interfaces that meet the diverse needs, preferences, and capabilities of the world's population. Customizable and diverse services and tools will also help us move toward true universal access and possible bridges for the digital divide.

We need interfaces that help us manage our public and private information flows. Tools that help us manage exoinformation (the information that leaves us consciously or unconsciously) will become increasingly important in both cyber and physical space. Today, our actions online are often collected and mined to serve our needs as well as to serve the goals of vendors or security agencies. Our smiles, gaits, and other mannerisms tell much about us to those who share our public and physical spaces. We typically welcome these inferences because they support mutually reciprocal relationships. The data gathering devices in the environment (video cameras, chemical and biological sensor networks) will make exoinformation in our physical spaces more accessible to others in non-reciprocal ways. One set of resources needed in the cyberinfrastructure focuses on helping us manage our privacy and make conscious decisions about carrying information across our private and public spaces. These resources must help us decide what aspects of our identities to reveal in specific contexts and subsequently remind us under what conditions we did so.

We need interfaces that help us manage memory augmentations—the logical extension of distributed digital libraries that are instantly accessible to us everywhere and anywhere. How we manage our distributed memories and our other cognitive augmenters is crucially important to health and productivity. When we are in high-performance cycles where our mental and cyber resources are in sync, these conditions should be sustained uninterrupted by the constant stream of

[28]The U.S. National Science Foundation has created a cross-directorate office of cyberinfrastructure to support research and development.

attention grabbers (phone calls, emails, spam, opportunistic alerts, etc.)[29]. Equally important, we need interfaces that protect us from overload and give us spaces to rest and reflect—cyber rest stops. Clearly, psychological and sociological research along these lines are needed.

All of these interfaces are broadly conceived to include a variety of resources and tools. They help us sort the information artifacts into public or private realms and they help us merge the cyber and physical spaces in innovative ways. Research and development focuses on the development of effective cyberinfrastructure that supports effective and efficient life in cyberspace. These examples lead us directly to the concept of information interaction in cyberspace.

6.2 HUMAN INFORMATION INTERACTION

It is ironic that the convergence of digital representation and electronic media in cyberspace are leading to vivid human experiences that are continual and contextual—i.e., analog. Furthermore, electronic information tools and methods decrease the distinctions between Voices1, 2, and 3. This blurring of the senses of information has led to an emergent phenomenon: information interaction. Electronic media and the notion of information as experience, in turn, lead to new senses of information. One that seems to be emerging is information as an extension of ourselves—our conscious and unconscious footprints and connections in the cyber world that bridge the physical and mental worlds.

Human information interaction is concerned with people and information as entities and the variety of actions that occur as people and information become proximate and reciprocally change one another. To understand human-information interaction more fully, we consider the characteristics of the actors (humans and information), the nature of the activities that take place, the kinds of relationships among acts and entities, the contexts in which interactions take place, and the resulting changes to the actors and the context[30].

6.2.1 USE AND INTERACTION

Throughout history, the artifacts humans have created were designed to respond to human initiation of use or interaction. Many of our artifacts are designed to cause changes in a third entity and little or no change in the human actor or the artifact. For example, we use a hammer to drive a nail into wood. In this case, we call these artifacts 'tools' and say that we 'use' these tools rather than interact with them. Some artifacts are designed to cause changes in the actor and little or no change in the artifact. We likewise 'use' these artifacts rather than interact with them. An automobile is a complex artifact that causes changes in our physical position, and we aim to minimize changes such as wear and tear on the automobile while it contemporaneously changes its position. We use our cars, initiate these uses, and during use there are sets of well-defined interactions with subcomponents of the car such as the steering wheel, accelerator, and brake that we initiate but also react to as part of the larger

[29]The program "Freedom" developed by Stutzman and Kramer-Duffield [2010] received considerable attention in the media because it allows people to disconnect their wireless connections for specific periods of time so they are not distracted by WWW interactions.
[30]See Marchionini, G. [2008] for a treatment of information interaction as a key thread of information science.

usage of the car[31]. Physical information artifacts like books are also 'used' in that we change (learn) as we use them but except for annotations and physical wear, the book remains unchanged. The 'use' of physical information artifacts is so important to human life that we have special terms such as 'read' a book or 'view' a film rather than the more general term 'use' a book or film. To describe the general use of multimedia artifacts, the term 'consume' is often used.

Taking a cue from cybernetics, we can consider the variety available to each interactor as an interaction parameter. Variety is used to measure the degree of control actors have in a systematically behaving (biological or mechanical) system (Ashby, W. [1956]). The amount of variety in actions available to actors (degrees of freedom) serves as one measure of use or interaction. This notion of variety in cybernetics is equivalent to Shannon's notion of choices in the communication source. When variety in the respective actors is strongly unbalanced (e.g., most of the control is in one actor), the activity tends toward use, and when the variety is balanced, the activity tends toward interaction. Furthermore, from a cybernetic perspective we can infer the following distinction between collaboration and use with interactivity as a measure. As an interaction tends toward giving equal variety to human interactors, the interaction tends toward collaboration; while when one interactor has more variety that another, the interaction tends toward use (coordination is a more polite term when referring to people).

Most of the artifacts created by humans are meant to be used; however, technology is increasingly leveraged to give our artifacts new properties that enable multiple and rapid interaction cycles and bi-directional initiatives, thus blending use and interaction. The alarm clock is a simple case where use begins to fade into interaction if there is a snooze alarm. Some children's toys have simple sensors and affectors that allow the toys to initiate interaction, for example, a toy that plays a recording 'play with me' one minute after movement has ceased. The modern built world is increasingly interactive. Terms such as 'intelligent' buildings, highways, automobiles, and homes actually mean infrastructural artifacts that may initiate actions and are available for interaction rather than only use. Thus, human artifacts and our constructed environment are becoming increasingly interactive[32]. We argue here that this is increasingly the case for information artifacts.

6.2.2 PEOPLE INTERACTING WITH INFORMATION ARTIFACTS

The most extreme example of people interacting with purely informational objects is immersive virtual reality. Virtual reality environments have grown increasingly 'real' over the past three decades. The sense of presence that is attained by today's immersive virtual environments is astounding. In highly realistic environments, the boundaries between the real and the artificial fully dissolve and

[31] Although cars that talk to us have not been successful in the marketplace, components that initiate action such as antilock brakes are well-accepted.

[32] Mitchell, W. [2003] Me++, presents an interesting explication of this phenomenon with many examples of how the built world is becoming extensions and appendages of our selves.

one's physiological reactions are as true in the artificial as in a real setting, so much so, that VR sickness is a serious concern[33].

Beyond the instances that mimic real life, there are important applications that enable people to experience dangerous (e.g., firefighter working in a flaming building) or physically impossible situations (e.g., walking around inside molecules and manipulating atoms). The applications of immersive virtual reality include entertainment (e.g., high-fidelity games), training (e.g., flight simulators), scientific exploration (e.g., physics and chemistry, astronomy, weather prediction), design and engineering (e.g., bridge and building design as well as simulated walk-throughs for clients), and measurement and evaluation (e.g., simulating nuclear testing in the post test ban treaty era).

Virtual reality is one example of how different classical senses of information apply in concert. Virtual reality is fully dependent on controlling stimuli (Voice2's communication process) to the senses. People in these environments are interacting with information in ways comparable to interacting with the physical world. The human actor changes (Voice1's thought and memory) as the experience progresses and the information artifacts change (Voice3) according to the human actions and planned conditions built into the simulation. When randomness or ambient contextual conditions (e.g., via sensors) are built in and when memory of previous interactions are taken into account (e.g., user patterns of past behavior), then the interaction is more dynamic and less predictable; i.e., more variety and information (Voice4's energy) is available. These immersive virtual reality environments are advanced examples of how people interact with information at very high fidelity, with very high levels of reciprocity and millisecond-scaled feedback cycles, giving rise to cognitive and affective senses of presence.

Presence is the term used to describe how 'real' is a mediated or vicarious experience. Lombard and Ditton [1997] provide a useful literature review of six different senses of presence: social richness (degree of personal intimacy), realism (fidelity, resolution), transportation (you are there, it is there, we are here), immersion (perceptual and psychological belief), social activity (degree of anthropomorphism, e.g., avatars with personalities), and social medium (degree that the medium is real to the perceiver, e.g., television or cyberspace). Many of the variables used to assess presence (e.g., time of engagement, fidelity, number of inputs and outputs, technical qualities, and user qualities such as degree of suspension of disbelief) are also useful in assessing interactivity.

Virtual reality aims to fool the human perceptual system into perceiving something that is not real. Human perception depends on rapid probabilistic assessments by the sensory organs. The different senses (e.g., visual, aural, vestibular, proprioceptive) have different thresholds for stimuli. The visual system detects motion at fine levels of difference whereas the vestibular system is less sensitive to very small motion changes. Virtual systems must therefore not only fool one sense system but the sum of the sensory inputs. We might consider the error in the human perceptual system Er_h (e.g., the amount of actual motion that a scene shows before the eye detects that there is motion) and the error in a virtual system Er_v (e.g., the fidelity and stability of head-mounted display or images

[33]Nausea and flicker vertigo are two of the more common side effects of immersive VR. *Washington Technology* ran a story as early as 1994 titled "Could the Surgeon General Warn: VR is hazardous to your health?" July 28, 1994 (vol. 9 no. 8).

projected on walls). When a person is engaged in a task, the Er_v is mitigated because the mental load (focus of attention) is on the task and many fine-grained anomalies are ignored. It is likely that technology will reach a point where $Er_v < Er_h$ for even casual immersions. In these situations, we will achieve very high levels of presence.

Less vivid but equally useful examples of information interaction are exemplified by work in telepresence, augmented reality, and by 2D virtual environments such as Second Life and the current popular attention to 3D movies and games. In teleprescence, a human controls remote devices (e.g., remote robot arms under the sea or in furnaces or biohazardous containers, scalpels in operating rooms) through prosthetics (e.g., a data glove in a safe laboratory with a high-speed connection to the robot device). In this case, information is a necessary mediator (energy flow) between the human and a physical artifact. The direct interaction from the actor's perspective is with the prosthetic system that presents visual and tactile information directly to the actor. With telepresence, we quite literally extend our physical reach across space.

In augmented reality, the surrounding stimuli are not suppressed, but they are dominant and overlayed with virtual information to enhance reality. For example, a surgeon wearing glasses that allow her to see a radiograph over the organ gains additional information without shifting vision from the organ to re-examine the diagnostic image. Less critical applications provide people with just-in-time information (e.g., names of approaching intersections in an automobile windshield) as they work and play. In both teleprescence and augmented reality, the information interaction is an integral part of the physical activity. In fact, the information interaction adds value to the human-physical object or human-human interactions. These cases offer yet more evidence for the ways that information and experience are becoming blurred in day-to-day life.

The realm of electronic games offers many examples of people interacting with purely informational spaces, either alone or with other people. Anyone who has tried to interrupt a youngster engaged in playing a video game knows how intense and 'real' are these information interactions. It is hard to imagine that such engagement is due only to the motor control over a joystick or other control device. In fact, the input and output devices are sometimes barriers to the full experience as the game progresses[34]. It is rather the case that the engagement is due to the mental linkage between the player's mental state and a set of bit states in the computer, i.e., a purely informational interaction. Turkle, S. [1995] was an early commentator on gaming communities and provides rich empirical evidence for the power that these information artifacts (games) hold for players both during play and in the lives of the players beyond the game.

A related phenomenon is the case of online environments or games where people who may or may not personally know each other compete or collaborate remotely to achieve ethereal goals. Dibbell, J. [1996] presents an early case study of serious emotional and legal issues that arose in an early Multiple User Dungeon (MUD) where one participant exhibited abusive virtual behavior, generating substantial community dynamics. The participant was subsequently exorcised from the

[34]Input/output devices that are designed for specific game conditions (e.g., steering wheel for driving games, drums and guitars for music games) have become popular ways to minimize the extrinsic load gaps between general purpose controllers and game objectives.

community (his access suspended—he was 'killed'), which, in turn, generated significant discussions about ethics, values, and laws in cyberspace[35]. These game environments allow participants to create virtual objects (informational objects) with rich descriptions and behavioral properties that are only limited by the creator's programming skill. People invest time and effort developing personas and rich environments. They become vested in communities that are composed of entities (both human and virtual) that may differ from the people behind the personas. These efforts to create personal projections and informational objects are so valued by people in these communities that they are bought and sold online using real-world currency.

The linkage between game world and physical world is becoming more blurred. The *Can You See Me Now* game created at the University of Nottingham (`www.mrl.nott.ac.uk`) involves multiple players, some of whom hide and seek in the real world and others play from their desks through a series of computer, GPS, and wireless connections. The experience is dependent on a series of information flows and displays, which players use for participation. See Bainbridge, W. [2010] in this lecture series for a socio-technical treatment of games. Bainbridge discusses what he calls the penumbra of the game universe where real life discussions and activities surround the in-game activities. The game (information interaction) is motivated as much by the social interactions in the physical world as much as by the actions in cyberspace and individual mental spaces. Other media are showing this penumbra effect as well. For example, the TV show 'Lost' (circa 2005-2010) generated many websites, blogs, and water-cooler discussions around the program.

Other kinds of information interaction are emerging through the combination of art and performance. Paul Kaiser's performance art piece called *Loops* uses finger movements by the dancer Merce Cunningham along with Cunningham's reading of his personal diary, and music composed by John Cage as database inputs to a series of adaptive computer programs that present a dance performance. Each performance (projected onto a free-hanging screen in a darkened theater) is unique and represents an ephemeral information act—just as a real dance performance is an ephemeral physical act. Not only is the 'information' never exactly the same twice, but it involves an interesting complex of human and machine interactions that yield novel products. In this case, the interactions are simulated and projected on-the-fly by information represented as computer code to present original interactions among the artists (dancer Cunningham, composer Cage, and choreographer Kaiser). Other performance artists aim to involve the audience more directly in the creation and execution of a performance, in some cases remotely.

6.2.3 AGENTS

Computational agents are programs devoted to performing specific kinds of actions that humans might otherwise take. There is a substantial amount of effort devoted to developing software agents (electronic information artifacts) that will act on our behalf in a variety of situations. Increasingly, we act with other people, objects, or information through information artifacts. We create and train these agents over time either explicitly or implicitly. In the explicit case, we consciously seed and tune

[35] Social environments like Facebook allow even more extensive aberrant behavior such as cyberbullying.

parameters, and add new values. In the implicit case, systems observe our behavior or the behavior of others and adapt the agent to these behaviors (e.g., recommender systems). In many implicit cases, we do not know that the agents even exist, as in the case of profiles that shopping services develop for us over time. Examples range from simple explicit information artifacts like 'out of office' replies that we set in our email when we go on vacation to increasingly sophisticated email spam filters that adapt to what we choose to view, and virus protection services that are updated over time as new viruses become known.

E-commerce agents aim to optimize shopping effectiveness and efficiency. Many online shopping services build profiles for us by recalling what we buy and relating these purchases to other shoppers' behaviors to make suggestions for us in our next visit. There is enormous angst and debate about these agents. They do have effects and will continue to become more sophisticated in influencing people's behavior.

There is considerable research in a variety of fields to understand and predict human behavior. Merchants employ search engine optimization companies to have their webpages rank highly in search engine results. For example, if you sell customized tee shirts, you are likely willing to devote a substantial portion of your advertising budget to having your company show up early in the results list for searches people do for 'tee shirts.' You may be willing to pay more to include search terms such as 'shirts' or 'causal wear.' If you are a small company with limited reach, you may be willing to pay for searches emanating from only specific location ranges. Search engine companies and other businesses use search analytics to promote sales and make marketing decisions[36]. Social psychologists study human reactions to a variety of physical and electronic stimuli. For example, a recent study demonstrates the strong influence that haptics make on decision making behaviors (Ackerman et al. [2010]).

Ultimately, consumers must become savvier about these techniques and deal with the trade offs they present. These techniques do offer some benefits to save us time and money on the one hand while stereotyping and homogenizing us on the other. For example, a variety of alerting services (e.g., RSS feeds) are emerging to help us stay abreast of personal or professional interests, and these will grow increasingly sophisticated in the future. When these become annoying is a personal response that is personality and context dependent. Usability of these services is dependent on two important considerations from an information interaction perspective. First, we must interact with the agents for them to be effective. This may be conscious or unconscious (biometrics may make these more pervasive and robust), but the interactions are real and potentially have important effects. Second, we must learn to manage these agents. This itself involves new kinds of interactions. Moving the spam filter I've trained on my PDA to a new office computer, for example, should be easy; however, this may not be the case.

These examples of information entities that are first class interactors illustrate a few of the possibilities and properties of human-information interaction. It is clear that each day brings yet another kind of information entity that includes some conditional properties, that acquires history

[36]See Jansen, B. [2009] in this lecture series for a treatment of search analytics.

that changes value over time, and is available to multiple people concurrently. These information entities interact with each other (typically at rates and volumes beyond those feasible for human interactions) to facilitate all manner of real world activity from banking to security to transportation to optimizing energy consumption in one's home. Furthermore, cheap connectivity and storage allow simple interactions to be stored and later aggregated and mined to facilitate new interactions.

There is a multiplier effect at play. One's spam filter gets better over time as will your oven, car, and other devices that store and pattern mine your interactions with them. Furthermore, devices can add data from other devices and other people to better develop services for you directly or as in the case of the online shopping industry for those who (optimistically) wish to better serve you or (pessimistically) separate you from your resources. These aggregated histories and profiles offer new sets of challenges related to property, privacy, and preservation.

6.2.4 LEVELS OF INTERACTION

Operational goals such as collaborating, competing, transforming, and observing drive the reciprocal actions that define interaction. People interact with other people or institutions in all of these ways. How do we interact with information toward these goals? Given an electronic information artifact, it is easy to see how people might observe and transform the information and how the information artifact can observe (monitor activity) and change (e.g., influence, inform) people. Competition is also easy to see if we consider computer games such as chess. Collaboration with others is strongly enhanced in cyberspace because space and time constraints can be relaxed (e.g., remote synchronized or unsynchronized communication), and interactions can be massively social rather than point-to-point (e.g., massively multiplayer online games such as *World of Warcraft*).

If we consider an information system such as an online travel agency, then it is plausible to say that people act on this service by specifying itineraries, and the information service acts on people by providing options and booking tickets. There is a further kind of coordination that may occur if the information service records my profile and saves me a bit of time each time I interact with it. There is even more coordination if all sales information is mined and the service offers recommendations (e.g., cheaper itineraries, special deals) based on these records. In this latter case, we have a series of interactions in which both the human actor and the information system are changed with each cycle of action.

Action is strategic. In the case of looking for a cheap and convenient travel itinerary, a strategic action may entail choosing a particular service to use (e.g., using a metasearch service) or organizing a set of queries that systematically examine different departure and arrival time alternatives. From the information entity perspective, strategic actions may consist of choosing which interface to present to a user depending on an IP address or personal profile. The strategic level also deals with establishing protocols to facilitate feedback in the interaction cycles.

At a finer level of organizational detail, we can consider tactical actions taken by actors and entities. These are purposeful acts that may be discretely described. For example, typing an airport code into a text box from the human actor side, caching a result set, and sending a properly formatted portion to the user from the information entity side of the interaction. The tactical level is especially important for human-information interaction research and development because it is the lowest level of conscious focus for most human interaction with information systems. Typing or uttering a statement or word is a common tactical action. People think and operate at tactical levels easily. It is tactical actions that lend precision to information interaction, and we use this level most often to discuss measures and properties of information interaction.

A still finer grained level of action is the move—a discernable act. Gestures, key presses or mouse clicks are examples of moves. Most human moves are executed by muscle memory. They are routine actions that serve tactics. Our verbal intonations, key press rhythms, and eye movements operate at the move level. To the human actor, moves are semi-automatic. A human or computational observer can collect these acts and use them to make inferences about the tactics and strategies that motivate them. From the information artifact perspective, displaying a page or accepting a request are moves. Optimizing move transactions is an important engineering and design problem. Part of the challenge of designing user interfaces is that machines are effective at the move level whereas people think and act at strategic and tactical levels.

There are, of course, finer-grained actions of interest within the actors (e.g., the various mental and physiological activities necessary to cause one's hand to click a mouse or one's voice to utter a word), and the information entities (e.g., the software and hardware actions necessary to compute results), receive and send data through various networks, manage databases, servers, and applications.

Typically, interaction involves all these levels of action, with the finer grained actions aggregating to comprise the more general ones. It is useful, however, to focus on one discrete act within the full interaction in order to develop well-defined features and measures of the overall interaction. By focusing on one act (e.g., tactical act), we can begin to characterize the amount of effort (e.g., time, resources) required for the initiator to execute the act and for the reactor to observe (e.g., understand) the act. We can specify the context for the action at different levels, most importantly, the goals of the actor(s) and the environmental and cultural settings in which the action takes place, and any preexisting relationships among the actors. More specific to the actors are the motivations and intentions behind the interaction. We can also specify the amount of information included in the act, as represented by the number of action choices the initiator has to choose from. None of these is sufficient, however, because interaction requires at least one round of feedback—a reciprocal act in response to the initiating act.

The act-react cycle is crucial to interaction and a single act-react cycle serves as the base for defining and measuring interaction. This reciprocal cycle can be considered at any of the levels of action granularity, but once it is specified, we are concerned with the time it takes to complete the cycle, the proportion of that total time that the different entities use, the nature of the act (e.g., physical, mental), and the degree of goal satisfaction the cycle yields. Thus, for a single interaction cycle, we can characterize features of the context, the participating entities, and one interaction cycle. An interaction that consists of a single interaction cycle is not very interactive in the ordinary sense of the term, and a new set of features and measures emerge as we expand to richer interaction where there are multiple interaction cycles. In fact, most people would say that process A is more interactive than process B if A offers more reciprocal cycles per unit time than B.

Fidelity and engagement may overwhelm simple rate of exchange. The number of cycles per unit time and the ratio of time to each participant are obviously important metrics to characterize interactions. The inter-cycle relationships are also important measures for interactivity. The depth of a discussion thread (the number of replies and responses on a given topic) is an indicator of interaction. Even more subtle is the degree to which the interaction tends toward continuity. This must be judged from the human point of view because electronic technologies can distinguish nanoseconds whereas people operate near the one second range. For the human, still images turn to motion images when they are displayed at roughly 100 millisecond intervals. Although it is possible to distinguish even smaller time intervals, it is reasonable to say that if an interaction proceeds with less than 500 milliseconds passing within any single reciprocal act, then the interaction will appear to be continuous to humans.

6.2.5 INFORMATION INTERACTION SUMMARY

Human information interaction is a disruptive phenomenon that changes the human condition. Throughout human history, we have primarily interacted with biological organisms, mainly people and animals. Electronic information artifacts present us with new possibilities and challenges. Reeves and Nass [1996] argue that our mammilian brains cannot adapt quickly to these new conditions and we have no choice but to treat new media like we treat people. This may be true at a gross level, but as human information interaction becomes more common to our everyday experience, we will surely adapt by adding layers of new tools that will help us make these differentiations.

We see that human information interaction can be measured in several dimensions: amount of variety available to the actors; degree of presence experienced; and number, frequency and fidelity of turn taking cycles. Increased levels of human information interaction are not only due to electronic information artifacts themselves but to people and their agents taking social actions in cyberspace that go beyond point-to-point personal interactions (e.g., Brown and Duguid [2000]). These developments have implications for personal and collective identities, a topic we turn to next.

6.3 PERSONAL AND PUBLIC IDENTITIES

Who are you? How do you think about yourself and your roles in the world? How do others recognize you? What do they think about your roles in the world? How do your actions in cyberspace influence the answers to these questions? These are some of the issues we consider in this section.

6.3.1 PERSONAL IDENTITY

Philosophers over time have pondered what we are and more specifically who am I, and am I the same person I was yesterday? Frankel et al. [2005] note that Aristotle proposed a 'substratum of essence' that remains even though our bodies change over time and that Locke, looking for empirical evidence of self proposed memory state as the essence of identity. Locke's notion that our personal identities are defined by what we remember has dominated identity theory and implies that identity is fundamentally informational (concepts in our heads ala Voice1 and artifacts we create to aid memory ala Voice3).

Like the notion 'information,' identity is a fundamental human concept that can be considered from many perspectives. Linguists (e.g., Muhlhausler and Harre [1990]) point to the importance of pronouns as an indicator of how fundamental identity is to language and thought (consider how many times the pronouns I, me, my, you, your, and our occur in everyday speech and writing). Hoffman et al. [1997] assembled a volume of papers that use history and culture to explicate how social identity influences personal identity. Our regional or national history and context influence how we behave, view ourselves, and are viewed by others. Hughes et al. [2006] brought together a set of papers that examine the medical, psychological, and legal dimensions of identity in dementia patients. Identity concepts also give rise to issues related to cloning (Kamm, F. [2005]) and technological self-transformation (e.g., steroids, plastic surgery). For example, Shapiro, M. [2005] considers the moral and legal identity issues of genetic engineering as people modify their own bodies and design their children.

Children learn to use pronouns and lose their ethnocentricity as they grow and have more life experiences. Our private identities (how we view ourselves) are shaped by actions taken and the reactions of others to those actions, especially families and friends. Young adolescents struggle to separate themselves from their parents and develop their own identities, developing ever larger circles of friends and acquaintances with whom they interact. Social media offer technological leverage to broaden the range of people who can affirm, mirror, and influence personal identities. Figure 6.2 depicts the mutually reinforcing effects of actions and products on identity formation.

As we live our lives, we act (chose what to wear, what events to go to, how to behave) and create products (work products, leisure products, and incidental products that leave traces of our acts). Farmers grow crops, construction workers build structures, and office workers create information products (artifacts) such as completed forms, reports, and messages. Figure 6.2 suggests that the sum of the actions and products shape our identities and that each act, product, and sum reflects back on our perception of self. The young man who thinks of himself as an astronaut but works as a farmer over time comes to reconceptualize himself as a farmer. In essence, we become what we do. There are two additional concepts in the figure that may strongly influence action: observations and records. The observations of others influence what we do and the records of our actions may limit or empower what we do. What we do not yet fully understand is how cyberspace actions and products influence our personal senses of identity.

6.3.2 PUBLIC IDENTITIES

As we live our lives, we develop public identities that are defined by our actions, the reactions of others to these actions, and records (information artifacts) of these actions and interactions. The clothes we wear, the way we shape our bodies (e.g., fitness, makeup), the words we say or write, and the actions

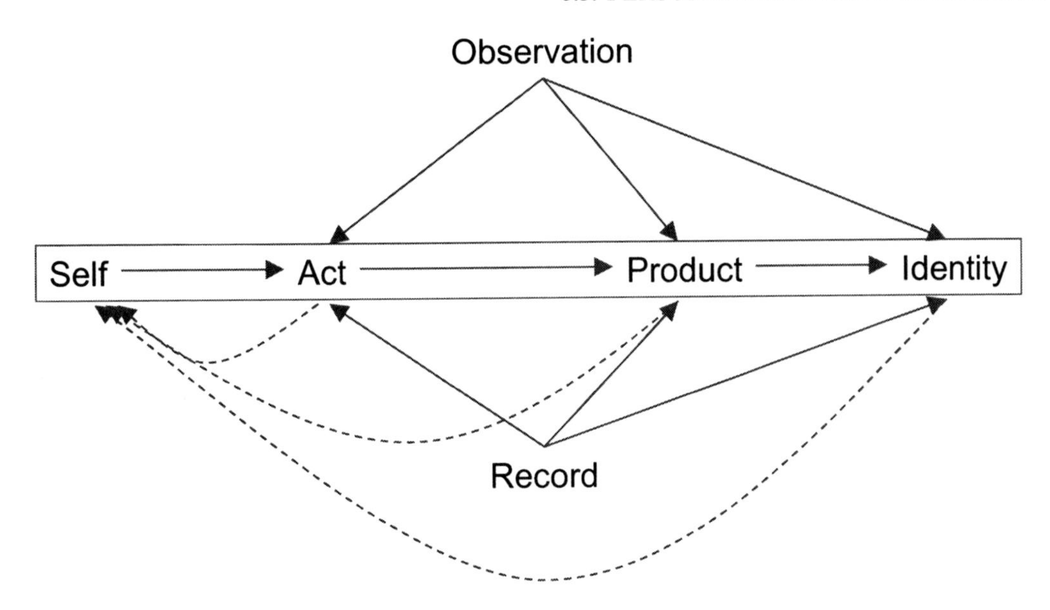

Figure 6.2: Personal Identity Formation.

we take all define our public identity, which is the basis for impressions that others form about us. Our public identities are important to others who assess, validate, and use our identities as the basis for interaction. From an observer's vantage, these public identities are what the observer thinks about us, and those thoughts are shaped by the actions and products we produce and whatever recorded histories are visible (e.g., pedigree). It is worth noting that our own sense of identity may be quite distinct from the perception of identity that others form of us. An entire public relations industry has developed around helping public figures shape the way others perceive their identity. Furthermore, because we often interact with strangers, societies create credentials that act as surrogates for personal observation of our identities (e.g., diplomas, licenses). All the relationships in Figure 6.2 are affected when we act in cyberspace. In cyberspace, information artifacts are easily shared broadly and new kinds of observations arise. Importantly, many of the artifacts of our interaction are created by automated processes over which we have little direct control. This is especially so for the observation and record processes that are extensive and mainly automatic in cyberspace. If all of the components of identity formation are effected in cyberspace, we should not be surprised that the resulting identities should be new phenomena.

Figure 6.3 illustrates the relationships between our private (personal) and public identities. Each of us develops a sense of identity that we hold in our minds as we live and this personal identity is represented in the figure by the leftmost stick figure. This mental image of oneself is what philosophers have tried to define over the ages. The public identity is represented by the large, fuzzy figure in the middle to suggest that it has many variants and is less consistent than one's personal

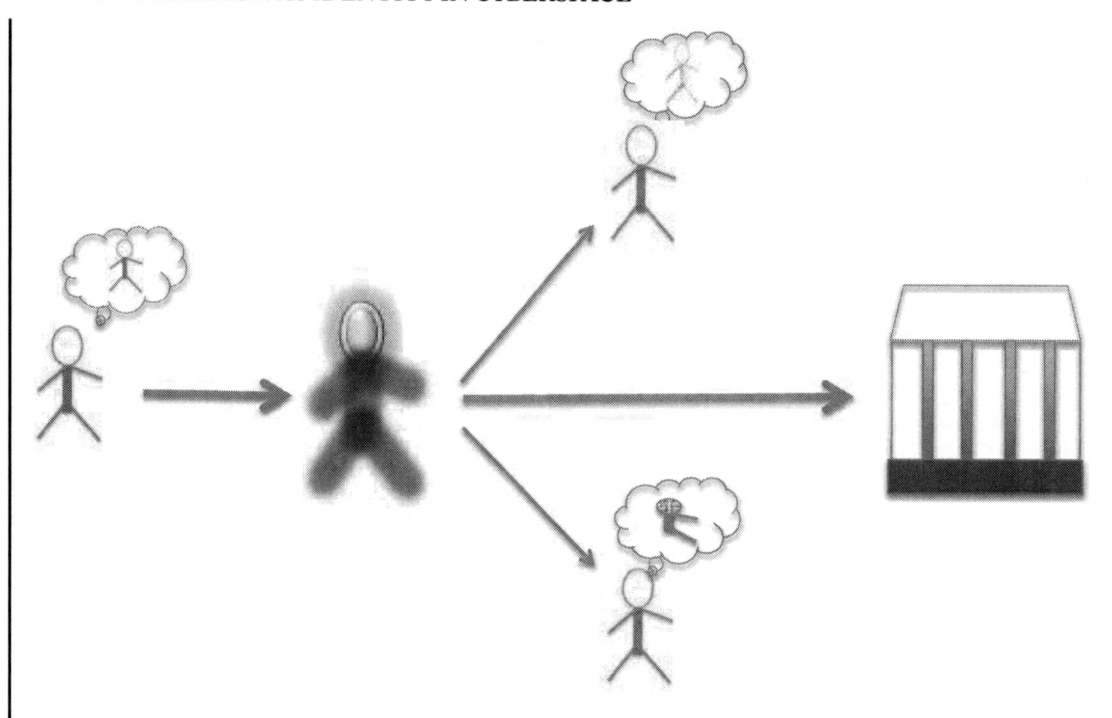

Figure 6.3: Private and Public Identities.

identity. The public identity is the basis for impressions of identity that others form. The mental images of us that others form are less well-defined and vary by individual. Thus, the images of us in the minds of others are shaded differently and with less definition in the figure. In preindustrial times, both our private and public identities were shaped by our local communities and culture and tended toward commonality rather than diversity. Cities, global travel, and more recently mass media support much more diverse public identities. This diversity carries with it the need to perceive, understand, and manage a broader collection of identities among the people one meets. This is a kind of cultural load we bear as we interact with more diversity. In cyberspace, this diversity and the associated burden of management is magnified. The building on the right in Figure 6.3 represents the institutional records of our identities that accrue over time. Traditionally, government agencies, religious institutions, and other cultural organizations have collected and validated these public identities. Automatic processes in cyberspace have opened up entirely new kinds of profiling of our identities by corporations and individuals—so much so, that these private and public identities are emerging as a new form of information that we turn to next.

6.4 PROJECTIONS, REFLECTIONS, AND PROFLECTIONS OF IDENTITY

Surely, we have always projected information to the physical and proximate world through our dress, posture, and behavior. Electronic media qualitatively change the scope of these projections in several ways. In cyberspace, we project ourselves in both deliberate and unconscious ways via information streams to others. The information projected is analyzed, transferred, saved, replicated, annotated, and interpreted by people and machines, many of which we have no knowledge about. As discussed in the previous section, every behavior in cyberspace is an information act that acquires history and 'pushes back' to shape and define us in both cyberspace and real space, i.e., reflects our identity.

People are highly skilled information producers and users. Our bodies and minds are tuned to gather information and to create it to serve our purposes. Each of Maslow's kinds of needs entail information requirements. Knowing where to get food, how to protect oneself from harm, and what will satisfy one's own or others' expectations requires both general purpose knowledge about the world and oneself as well as context-specific information. In the previous chapters, there were many examples of information coming into our beings—we seek it, filter it, assimilate it, and use it to make decisions. As information producers and consumers, we also have substantial streams of information that leave us as we live our lives. Some of this information is purposeful and conscious (e.g., messages to others, publications, webpages, blog postings) and much of it is not (e.g., facial expressions, gestures, keystrokes, click streams). This is exoinformation (Brunk, B. [2001]), which serves as *projections* of our identity.

Additionally, people may create artifacts about us (e.g., comments on our projections, tags or links to our artifacts) and computational agents may also create artifacts about us (e.g., spam filters, purchasing profiles). These artifacts created by other people or agents are *reflections* of our identities. Taken together, our *pro*jections and *reflections* represent *proflections* of self and represent a new sense of information.

Figure 6.4 summarizes these concepts and estimates respective volumes of information artifacts of each type. Note that the estimates are guesses at this point and cry out for empirical data. This is a research challenge on several fronts. The figure assumes equal amounts of purposeful and incidental artifacts although this is open to debate. Our projections may be purposeful (e.g., a personal webpage or a social media profile) or incidental (e.g., the time or place automatically associated with a mobile device action). In either case, we may be or become aware of the status of these projections. For example, an email message we project may be forwarded to others with or without our knowledge, and in the latter case, we may eventually find out about the forwarding or not. Presumably, we are aware of most of our purposeful information streams; however, we may forget or set automatic feeds that accumulate over time. We are unaware of most of our incidental streams. Our reflections may be created by people or by computational agents. The figure suggests far more reflections by computational agents than humans. We tend to be unaware of our reflections (when we do a web search for our names or mentions in a microblog, we aim to discover some of these reflections).

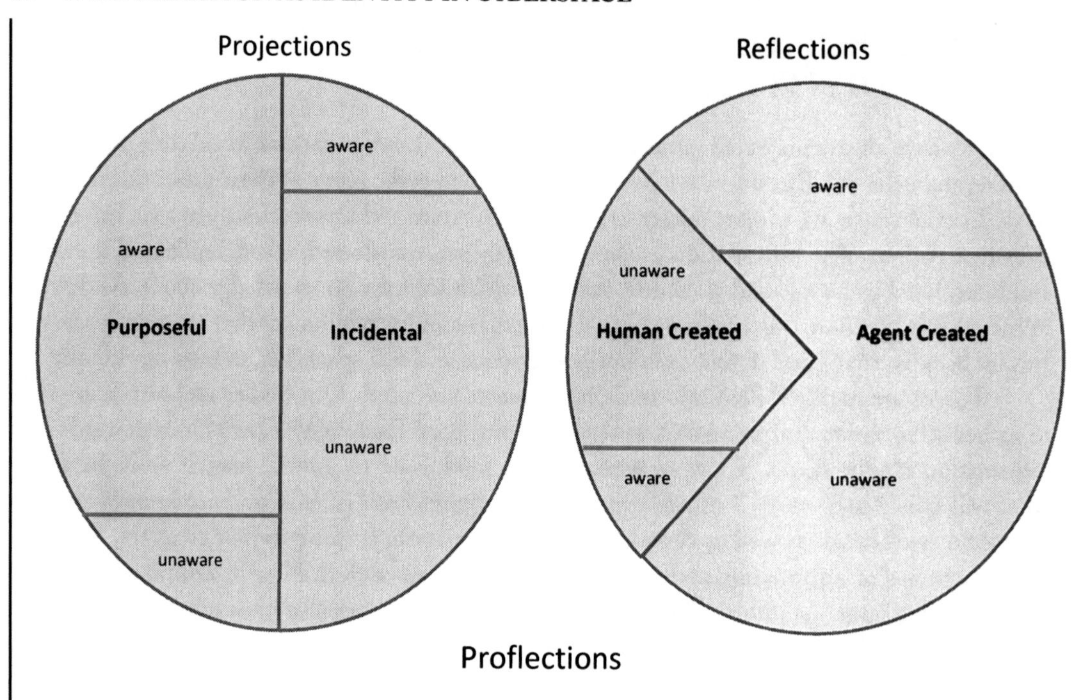

Figure 6.4: Projections+Reflections=Proflections.

6.4.1 EXOINFORMATION AND PROJECTIONS

Our words and actions may be observed by others and subsequently interpreted with consequence. Our thoughts are not directly observable, so information as a state of mind is outside the realm of direct exoinformation, except so far as inferences can be made about those states based on our behaviors. If I walk into a restaurant dressed in expensive clothing, the waiter makes some quick observations (gathers information) and takes actions according to his inferences about me based on these observations. However, if I go to the same restaurant in ragged clothes, the observations may be more extensive and the subsequent actions quite different. I am the same person (my private identity is the same); however, the information that I project with my clothes and manner of behavior may have strong effects in the world. In fact, humans are continually projecting information into the world in conscious and unconscious ways. Experienced adults are aware of much of the unconscious information we project, and we work to control that flow of information to influence what others think about us. Thus, we selectively push out streams just as we selectively attend to incoming streams. The more public our projections, the more careful we are to 'act' in ways that shape our public identities. One extreme effect of this selective projection is called political correctness.

Interacting with information in cyberspace is doubly selective: we filter our actions as well as our sources. These interactions change us and they change cyberspace. A challenge in cyberspace is that we are often alone in physical space or even if we are in a public space (e.g., a coffee shop), our focus is on a semi-private screen and keyboard (or PDA/cell phone), and we tend to act as if we are in private space because many of the public cues are ignored or suppressed. Thus, our perceptual system senses privacy when in fact we are acting in massively public space. In cyberspace, our in and out selective control streams are unusual to our typical experience and lead to emergent conditions that we are only beginning to understand.

How much time do you spend in cyberspace each day? A significant result of electronic technologies is that much of our daily activity is done in cyberspace and we have not learned how to 'act' in this new environment. This is highly problematic because we do not have mutual awareness of all the possible observers in cyberspace. Walking down a busy street, there are many people who may observe us and we act accordingly. In cyberspace, the physical cues from the environment are impoverished, and we have small gauge on how much control to exercise on our exoinformation. Because there are so many potential observers, we must assume that we are acting for the entire world, but this is artificial to our day-to-day experience. In fairly low-fidelity information channels such as email, this is relatively easy to do (although most people can recall the embarrassment of sending a personal email meant for a single individual to an entire electronic list). Your emails or text messages are small 'chunks' of you and may be saved or forwarded by others with or without your knowledge—they may become part of your 'brand' or a kind of pheromone of your behavior. As cyberspace becomes more rich and high-fidelity (e.g., webcams, location-aware sensors, and time stamps), understanding what we have projected and to whom will become increasingly difficult. Not only do we not know who is observing our exoinformation, but our observable life traces (exoinformation) are becoming richer as they include not only our words and click streams but our temporally linked locations and physical states.

As if our lack of cues about online human observers was not challenge enough, the problem is exacerbated by the rapid growth of sensors in the environment and machines programmed to monitor our online behaviors. In the grocery store, I am not surprised that a clerk observes my exoinformation projected by my actions and asks if I need assistance finding something. If the clerk follows me around the store and continually asks if he can be of service, I am likely to become annoyed. When I check out of the store, I make a conscious decision to allow my grocery store to observe specific exoinformation (what items I buy by scanning my magnetic identification card) in trade for some cost savings. This exoinformation collected by the store as I pay for items is valuable in several ways. First, it allows the store to better control inventory by knowing exactly what should be reordered, and when aggregating my exoinformation with other shoppers, what products are selling at different times. Second, it allows the store to target me for particular products by sending me sales information (and sell this information to product manufacturers who can also target me for these or related products). Today, grocers augment their grocery sales by selling people's exoinformation to manufacturers and distributors. This example demonstrates cases of projection for which I have considerable awareness (the clerk following me) and for which I have little awareness (the specifics of the profile that generates coupons and targeted ads for me).

Beyond the direct targeting for the products I buy, businesses can develop sophisticated profiles of me as an individual to then offer other products I might be willing to buy. These profiles are important examples of reflections of me that will be treated in the next section. They can also mine the aggregated data to see patterns for introducing new products or reorganizing shelf placement to optimize sales. For example, by knowing that many customers that buy product A also buy product B, I can be reminded to buy B each time I buy A, or product B and A can be grouped together in physical space or hyperlinked in cyberspace. The field of data mining is burgeoning and product sales are one important application. Although this trend toward using personal actions to develop profiles and aggregate actions to optimize sales has been widely debated and analyzed in the scholarly and popular press, people seem quite willing to trade their exoinformation for cost savings or better service even if we have little or no control over the resulting reflection. We have not quite gotten to the point where we simply send an automated shopping cart down the aisle and the products we 'want' jump off the shelf into the cart automatically, but profile-based advertising is pervasive and the technology and our knowledge of human behavior are both getting better[37].

Today, these are blanket trades we make for our exoinformation since making individual decisions every minute is unacceptable. In fact, better targeting of information to specific people at specific times and locations (often called just-in-time) is a win-win for sender and receiver. It helps the sender to increase the chances of effectiveness (just as the yellow pages is a better investment for merchants than newspaper ads because people who consult yellow pages are already predisposed to buy) and helps the receiver by providing presumably pertinent information AND reducing the amount of unwanted information (spam) that might otherwise be received. The benefits depend on well-constructed profiles. There are also situations in which exploration and browsing are what is desired and profiles should be dampened under our control.

There, is of course, a darker side with respect to privacy. In many cases, these trades are not conscious at all since much of what is considered to be public behavior is recorded and mined without our knowledge, or more precisely, we do not take the time to read terms of service agreements. Consider the agreements you make when you click 'agree' to a long list of legalese as you begin to use or upgrade some online service. Most of these agreements allow use of your projections and even when you are permitted to disallow this, it is often via an opt out rather than opt in selection. Beyond our willingness to consciously trade exoinformation (personally or through our electronic agents), there are increasing numbers of systems that gather such information as part of the built environment. Electronic health records and personal health records are increasingly promoted as ways to insure persistent and accurate bases for diagnosis, treatment, and patient safety. Security and safety concerns have led to video cameras in many places. When combined with facial recognition software, our very presence in those places leaves traces (projections) that we are likely unaware of. Additionally, sophisticated sensor networks that gather biological, chemical, and other data continually are emerging which will provide additional projections for future mining. The legal

[37] Radio ads as you get near a store or restaurant, personalized verbal greetings and sales advice when you walk into a store, sales or service information projected onto one's augmented glasses as one walks the streets.

rights to restrict data access are somewhat stronger for these systems; however, data breaches are common.

Projections become much more crucial in electronic environments because we project ourselves into many virtual places, possibly simultaneously[38]. The extreme case is identity theft where others project reflections of us with possibly devastating consequences. Less extreme, but perhaps more common, are the cases where others can use technology to examine our projections without our knowledge. Many social media sites allow anyone to examine profiles or probe our social networks. Sometimes, we can restrict such probes (Facebook has changed its privacy settings several times and in late 2010 allowed such restrictions to be explicitly set, although the default remains open to all). Twitter feeds, at this time, are fully open and harvestable by a variety of tools and institutions (e.g., the Library of Congress is preserving all Twitter postings)[39].

In the physical world, people have learned to manage their projections by filtering their behavior. In cyberspace, these projections are more pervasive, global, and dangerous because there are so many human and machine observers, and we have no way to know they exist and what are their intentions. Thus, there is a growing new sense of information that arises from our increasing experiences in cyberspace. Given that it is possible to have ongoing relationships with people we never physically meet, and furthermore that there are people and machines that have one-way relationships with us without our knowledge, it is the case that our identities *are* information and these projections themselves are a new sense of information.

6.4.2 REFLECTIONS

We are not alone in the physical world. Cyberspace is even more heavily inhabited because there are computational agents as well as people, and these agents are charged with observing our behavior. We send or leak information, and other people create information about us. This information becomes part of our identity. Biographies and news updates have long been common for famous people. In the past, most people had to settle for the snarky notes written about us by our middle school friends. Cyberspace democratizes public relations. Everyone who works or plays in cyberspace is subject to a steady stream of comments, tags, annotations, and links to our projections as well as original postings about us that are not associated with specific projections. These additions to our projections and original postings about us, whether human or machine created, are reflections of our identities.

Figure 6.5 illustrates one set of reflections that exist in the scholarly publishing world. Scholars cite the work of others, and these citations illustrate a kind of relationship. Indexes of citations, co-citations and other patterns are created to assess impact and to discover influences. Cyberspace makes it easy to find and analyze these same kind of relationships in social networks of friends rather than authors citing each other.

[38]In the virtual environment, Second Life, people create avatars that represent themselves in different cyber settings and one may create robo-avatars that follow or act in place of your primary avatar.

[39]Services like TweetStats (`www.tweetstats.com`) allow anyone to view tweet patterns for anyone (e.g., by year, time of day, frequency, etc.).

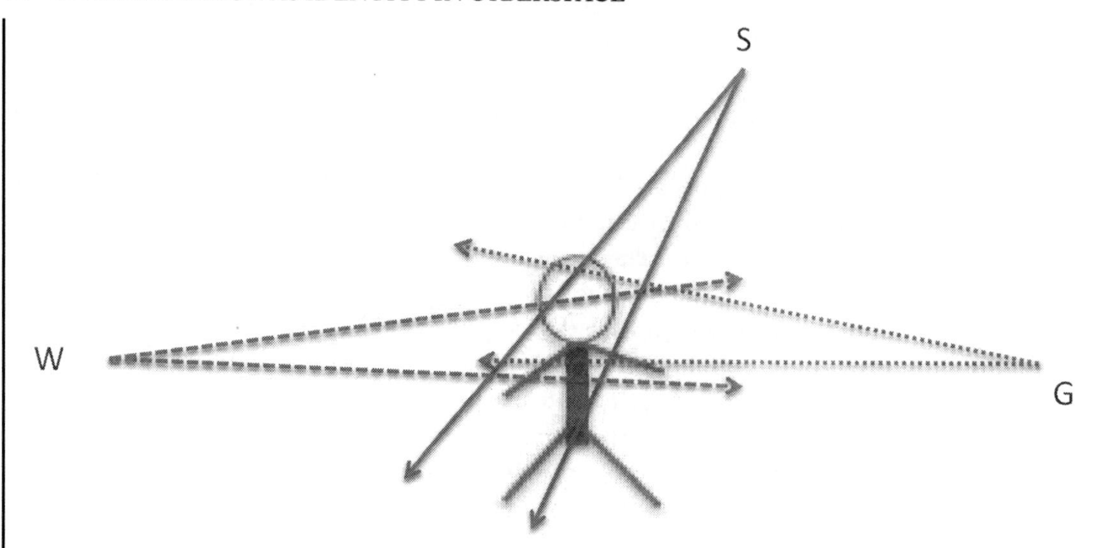

Figure 6.5: Three reflections of publishing identity.

In Figure 6.5, the stick figure represents all the publications (projections) by an author. As noted in Chapter 3, services create citation indexes for authors and these indexes are reflections of author identities. For example, one citation service will provide a profile with number of publications, citations to those publications, and various indexes of impact (e.g., h-index). In the figure, the three letters stand for different citation indexing services (e.g., Web of Science, Scopus, Google) that each use their own proprietary techniques to find all publications by an author and citations to these publications. Each service will have a slightly different view of the author's citation influence (see Meho and Sugimoto [2009] for an example of how different services compare). Thus, each service has a different representation (reflection) for this aspect of an author's identity.

Now consider a social network of Facebook friends. If I have 100 friends, each one has a slightly different view of me. These views have some common elements (e.g., my profile and my public 'wall'), some have fully private views (e.g., the private messages we have exchanged), and some have partial commonality (e.g., some friends in common, group interactions). Thus, each of these 100 friends has a unique view of me. A specific friend will have the projections of self that I make as well as the public reflections of my 99 friends and his/her own private reflections on me. At present, no single person has access to the 100 unique views of my identity and social media services are careful not to allow general-purpose harvesting of these proflections.

6.4.3 PROFLECTIONS

Taken together, the totality of projections and reflections in cyberspace define an individual's proflection of identity. Of course, the totality is never visible to anyone, including ourselves because no single entity has full access to all of cyberspace at once. Additionally, the totality is changing over time, even beyond our physical life. Thus, at any given instant, what we or others can obtain is a view of our proflection. These views are a new kind of information. They are not information in a human mind ala Voice1 but rather in the cyberspace collective. They are not one person or group informing another ala Voice2 but emergent, interacting social forces. They are not static artifacts but dynamic interactants.

Consider the Voice1 point of view that information is strictly in the head as memories and thoughts. Human nature is such that some of our behavior is automatic and some is driven by our conscious thought—by the Voice1 kind of information. Although most people believe that behavior is driven by conscious thought, Bargh and Chartrand [1999] summarize empirical evidence to argue that the bulk of human cognitive activity and behavior are driven by nonconscious, automatic mechanisms rather than intentional, conscious control. Additional arguments that challenge conscious choice as the seat of behavior are made by neuroeconomics that combines social-psychology studies of decision making and neuroscience data from brain imaging (see Glimcher et al. [2009] for a collection of readings). Regardless of whether behavior is conscious or automatic, it can be observed and inferences made based on those observations. Psychologists and anthropologists construct models for internal mental processes such as motivation, reasoning, emotion, and learning based upon observing human behavior in different settings. Many of these techniques and models are being applied to our behavior in cyberspace.

As more and more of our behaviors take place in concert with machines and in cyberspace, there are a host of new kinds of electronic observers that create models or profiles for us. They include hardware sensors, including some we take onto and into our bodies to monitor physiological processes, as well as software processes that pattern match and make probabilistic inferences. Thus, in the world external to our heads, a large number of profiles of us are created that aim to predict our actions before we decide to take them. These external systems analyze the information artifacts we produce, use the traces of our behavior and instantaneous states of our bodies, along with the aggregate of other people's behaviors to make inferences about us. There is considerable concern about this growing trend to predict human behavior. Lanier, J. [2010] argues on several fronts that this trend is dangerous. He notes that data mining reduces our humanity, and that uncontrolled reuse leads to mediocrity and stiffles creativity. Such critiques are increasingly common and illustrate that we are indeed beginning to consider the implications of life in cyberspace.

Let us distinguish our memories and our histories. Our memories are the traces of our mental, spiritual, and physical lives that we create and control. Our histories are the traces of our lives that are captured and controlled by us and by the environment, including other people. Our memories may be strictly internal to us (mental) in the sense of Voice1 or they may be externally supplemented by artifacts (e.g., writing, recordings, photographs) as in the information of Voice3. Traditionally, histories are artifactual in nature with the explicit linkages among these artifacts a kind of mirror of our lives, and the implicit or unrevealed linkages among these artifacts might be thought of as a kind of shadow of our lives. It appears that social psychology and technology advances leverage history to infer memory and to predict behavior.

In addition to the memory-history distinction, consider three levels of information sharing: private, semi-public, and public. Private information is meant to not be shared. It should not be projected. However, for a variety of reasons, what we intend to be private may be shared or it may be projected as exoinformation. Semi-public information is meant for a specific set of people or systems. Collaborative work in an organization is shared within the group but not with the world at large. Internal memos and Intranets are meant to support this kind of information sharing. Public information is meant to be shared with everyone. Publications and the open WWW supports this kind of information sharing.

Using the private-semi-public-public levels of sharing intention as a dimension, we can depict different kinds of relationships between memory and history, conscious and unconscious behavior, and physical and electronic information artifacts. From an exoinformation point of view, we might characterize the amount of information leakage when people work with physical or digital artifacts. The amount of exoinformation that leaks from us is much higher when we work with digital artifacts[40]. For example, libraries in institutions that have classified materials are concerned with protecting user actions such as simply reading a paper from outside sources. When journal subscriptions are physical, this is easy to do, when the journals are served from the publisher systems, it may be necessary to locally mount the database so that individual reading behaviors cannot be tracked outside the institution. Semi-public information in digital artifacts tends to move into the public arena easily as demonstrated by employees who are fired after discussing their work on blogs or posting to other social media. It may be argued that private information is impossible in cyberspace. Others may argue that private cyberspace compartments will emerge as new kinds of services just as banks did in physical space[41].

Consider the private-public dimension crossed with memory and history with respect to how working with physical and digital artifacts blurs the boundaries between the private-public. The private-memory cell is easy to make semi-public or public with digital information artifacts, but requires much greater cost and effort with physical information artifacts. The amount of effort acts as an impedance on the flow of private to public, and this impedance is quite low in cyberspace.

Our proflections serves as our image or brand in the increasingly important cyberspace world that has real impacts on our lives. The main issues are the following: privacy, personal information management, preservation, personal success in an information society, and associated self-actualization.

Privacy is an obvious issue as multiple kinds of projections leave us consciously and unconsciously, and people we may or may not know reflect us via links, reposts, and annotations. It is a much more pervasive issue in cyberspace where countless people, machines, and software systematically observe our information behavior with our knowledge and more often without it. The European

[40] One of the cues to British and American physicists that German physicists were working on atomic weapons was when they stopped publishing related topics in the physics literature; a related clue is to observe what leading physicists read.

[41] Hartzog, W. [2009] proposed a 'privacy box' for social media sites that depends on participants signing a confidentiality agreement.

Union has published guidelines to protect people's privacy in environments where computational resources are transparent[42] and various privacy seals have been developed by consumer and business groups for e-commerce applications, although Moores, T. [2005] argues that few people understand them. In cyberspace, thousands of programs and agents make inferences about us based on our online behaviors at a much larger scale than the number of people we could possibly meet in the physical world on a typical day.

Information management has moved from the library and archive setting to personal offices and homes. People with moderate income resources can easily accumulate thousands of high-resolution photographs, scores of hours of digital video, thousands of music files, and thousands of emails and other text documents each year. The information retrieval community has traditionally focused on large-scale public or corporate information artifacts; however, it now also races to keep up with the challenges of personal information management. Given the amount of exoinformation collected by the various systems we interact with today, it might be easier for some third party to access our personal information artifacts than for us to do it ourselves. Whether we have any formal relationships with these personal librarians (mostly in the form of software agents) is partly a privacy issue and partly an economic issue.

The information management issue leads directly to the issue of preservation. Given the rapid evolution of artifact formats and the hardware necessary to use them, we not only are challenged to find our external memories and what histories the environment has compiled, but also to preserve them. Many people have stacks of floppy disks, removable disks, and memory chips, cards, keys, and other devices in various sizes and formats that are no longer usable. Information artifacts that were traditionally considered ephemeral are now routinely saved, backed up, and archived (three different kinds of operations) because it takes less time to do this with electronic artifacts than to make decisions about whether to keep them or not. This leads to redundancies and more costly searches in future usage. Furthermore, this requires ongoing vigilance and migration as improved coding schemes and hardware systems emerge.

Finally, our ability to work in cyberspace and manage our proflections will determine our professional success in an information society. This is so because we potentially reach a larger portion of the human population and an entirely new population of non-human systems when we spend so much of our life online. The impression that most of these people and systems have of us are views of our proflections. These impressions, in turn, affect us through various feedback loops that determine our success and eventually our self images.

My proflections in cyberspace are partly mental, partly acts—frozen behaviors and potential actions, and partly artifacts that are fluid brands for my identity. Although I may personally view myself with a consistent self-identity, my proflections are multifaceted identities that different people view through different lenses that are defined by their own proflections, local contexts, and the instantaneous state of cyberspace instantiated by their technological viewpoint.

[42]European Disappearing Computer Privacy Design Guidelines V1.0, Ambient Agora Report D15.4 (October 2003).

CHAPTER 7

Conclusion and Directions

In this lecture, we have considered several different classical senses of information, with particular emphasis on the ways that electronic systems are influencing those senses and give rise to new phenomena such as cyberspace, information interaction, and new senses of information itself. The evolution of cyberspace as an informational place between our physical and mental spaces offers new opportunities for human collaboration and problem solving as well as new challenges to learn, work, and play safely, and with civility.

The incorporation of programs with conditionals sensitive to context from human actions or sensors embedded in information artifacts makes interaction with those artifacts possible and increasingly common. This human-information interaction causes a fundamental shift in how we think about information. Information becomes more experiential, fluid, and massively social. Information in the head (Voice1) becomes externalized as more external memory prosthetics are adopted and as sensors and analytic engines make inferences about our intentions. Information as act (Voice2) becomes persistent rather than ephemeral as our digital actions are echoed and amplified in cyberspace. Information artifacts (Voice3) become dynamic extensions of ourselves and our culture and acquire layers of context that reverberate with human and machine usage over time. The general sense of information as energy (Voice4) is amplified to global scales.

These changes and convergences in the classic senses of information are joined by new kinds of information in cyberspace. One of the emergent effects of these changes is the way that our personal identities are projected and reflected in cyberspace. The resultant proflections of self represent a new kind of information that gives rise to new uses by institutions, computational agents, and other people (i.e., the fodder for decisions) and that we must learn to understand and manage. From the personal perspective, these extensions of information give us new powers but also bring new responsibilities for behavior.

What do these changes in information mean to our lives? At one practical level, they are disruptive to existing information industries. The global information industry has been revolutionized in the past two decades as publishing, broadcasting, and telecommunications have been augmented by the Internet. Billion dollar software and search engine companies and social media services have arisen while traditional information services such as newspapers, libraries, and media studios adapt to use new tools and to offer new services. These trends are certain to continue for the foreseeable future as human creativity leverages cyberspace and interactive information artifacts to invent new services and products.

Immediate access to information must also affect how we value information. In place of scarcity as a driver of value, accuracy and trust become the linchpins of value. Open source, open access public information enables cyberinfrastructure (e.g., Linux, Apache) and gives rise to new collaborative information sources (e.g., Wikipedia). These developments are championed as evolutions of collective intelligence on the one hand (e.g., Shirky, C. [2010] arguments for technology-enabled human generosity with free time supporting a contribution culture to supplement our consumption culture) and questioned as mufflers on creativity and economic motivation on the other (e.g., Lanier, J. [2010] concerns that massive reuse devalues creative effort) with a variety of compromises for hybrid evolutions in between (e.g., Lessig, L. [2008] examples of how creative reuse generates new economic models). These debates demonstrate welcome maturity in cyberspace as implications for people become as discussed and promoted as technical progress. Questions of intellectual property eventually lead to questions about the nature of expertise and trust. Students with PDAs in classes are able to challenge their teachers, facts can be challenged as opinions, and expert opinions can be countered as they are consumed, annotated, and reused. These trends can be positive or negative depending on the whos and whys involved. What is clear is that accuracy, the chains of evidence that make accuracy transparent, and trust take on increasingly important roles in cyberspace.

Our educational enterprises (both self-directed and formal system) must evolve to prepare us to live in cyberspace as well as in mental and physical space. Gathering, filtering, and analyzing data become essential and complex antecedents to creating and sharing new ideas. Techniques and tools for collaborating and using globally scaled data sets will migrate earlier in our educational experiences. Additionally, knowing how to remix and reuse information artifacts is a new kind of literacy. This will require new instructional strategies and ways to assess learning outcomes. Our formal education institutions are joined by a variety of public and private entities that do not depend on synchronous or co-present learners and teachers. Perhaps more significantly, self-directed learning and creative uses of cyberspace in all kinds of formal learning institutions will generate new demands for systematically organized and trusted digital libraries.

At the system level, we must create services and user interfaces that help people manage their proflections. One class of tool might be thought of as digital condoms that help us protect ourselves in cyberspace from leaving and receiving artifacts not pertinent to our goals[43]. We need suites of tools that remind us about information streams that leave us through routine use, that help us manage our privacy settings across different applications, and that help us distinguish physical, cyber, and mental spaces. Of course, these tools and services must be easy to use and not themselves become new burdens to manage. Tools and services (and perhaps laws) that change our proflections will be more difficult to create. Clearly, it will be easier to manage our projections than to edit our proflections[44]. In addition to tools and services, we need methods and models to study proflections. We require techniques to combine the statistical tools of social network analysis together with ethnographic strategies to advance our understanding of personal identity in cyberspace.

[43] OpenID (`http://openid.net/`) is one example of a centralized service to manage passwords and profiles rather than maintaining individual settings for different web services.

[44] Russell and Stutzman created ClaimID (`http://claimid.com/`) as a service to help people manage their online identities by specifying which web pages were pertinent to them and which were not.

We cannot depend on technology alone to manage these new types of information. At the individual level, as we spend more time in cyberspace, we must learn to take more personal responsibility for our actions and be more tolerant about the reflections others make upon us. We must develop more nuanced senses of civility: a snarky comment made to a friend in a face-to-face setting rich with context that makes it funny does not scale to cyberspace where both the rich context is missing and a plethora of others can consume and echo the comment. Just as tremendous good can come from collective action marshalled by social media, hate can also be amplified dramatically if we are not aware of the what boyd, d. [2007] calls mediated publics that reverberate and persist long after the emotional heat subsides.

Can we influence our proflections? Does it matter how people and machines perceive us in cyberspace? Concerns with cyberbullying and identity theft suggest that it does matter. Clearly, we must learn how to better manage our projections, and there will likely be legal, technical, and social 'thermostats' created to help correct or mitigate reflections. For example, 'human flesh search engines' leverage connected people to both unite groups for collective action or to punish bad behavior in both physical and cyber worlds (Wang et al. [2010]).

More generally, we each must become more mindful of how we want to shape the future. How much should our children learn from simulations in cyberspace and how much must be rooted in the visceral physical world? What ideas and information artifacts should we own or at least demand attribution? What are the proper rewards for creating new ideas, information artifacts and services? Do we shelter ourselves from cyberspace (curtail our projections and ignore our reflections) until privacy-preserving techniques and systems emerge? Or do we rather embrace the notion that privacy is an outdated concept and accept total openness to our behaviors and our minds? Do we accept full tolerance of any behavior in cyberspace or are there principles and laws that must obtain? Do we embrace the mining of massive social and personal sensor data to develop new neuro-economic and socio-political strategies to optimize and control human behavior, or do we champion the heroic rebel running against the emerging winds of cyberspace? There are no simple or correct answers to these questions, but we will face them directly or ignore them, and the future of our progeny will live with the results of our in/actions.

Bibliography

Ackerman, J., Nocera, C., and Bargh, J. (2010). Incidental haptic sensations influence social judgments and decisions. *Science*, 328, June 25, 2010. 1712–1715. DOI: 10.1126/science.1189993 60

Agarwal, S. Furukawa, Y., Snavely, N., Curless, B., Seitz, S., and Szeliski, R. (2010), Reconstructing Rome. *Computer*, 43(6), 40–47. DOI: 10.1109/MC.2010.175 43

Alavi, M. and Leidner, D. (2001). Review: Knowledge management and knowledge management systems: "Conceptual foundations and research issues". *MIS Quarterly* 25 (1): 107–136. DOI: 10.2307/3250961 8

Anderson, J., Carter, C., Fincham, J., Qin, Y., Ravizza, S., and Rosenberg-Lee, M. (2008). Using fMRI to Test Models of Complex Cognition. *Cognitive Science: A Multidisciplinary Journal*, 32:8, 1323–1348. DOI: 10.1080/03640210802451588 15

Applebaum, D. (2008). *Probability and information: An integrated approach* (2nd Ed.). Cambridge, UK: Cambridge University Press. DOI: 10.1017/CBO9780511755262 47

Ashby, W.R. (1956). *An introduction to cybernetics*. London: Chapman and Hall. 56

Bainbridge, W. (2010). Online Multiplayer Games. Synthesis Lectures on Information Concepts, Retrieval, and Services. ed. Gary Marchionini, no. 11. San Rafael: Morgan and Claypool Publishers. DOI: 10.2200/S00232ED1V01Y200912ICR013 59

Bargh, J. and Chartrand, T. (1999). The unbearable automaticity of being. *American Psychologist*, 54(7), 462–479. DOI: 10.1037/0003-066X.54.7.462 73

Bell, G. and Gemmell, J (2009). *Total recall: How the E-memory revolution will change everything.* NY: Dutton. 33

Berners-Lee, T., Hendler, J., and Lassila, O. (2001). The semantic web. *Scientific American*, (May, 2001). DOI: 10.1038/scientificamerican0501-34 15

Bjorneborn, L. and Ingwersen P. (2004). Toward a basic framework for webometrics. *Journal of the American Society for Information Science and Technology*, 55(14), 1216–1227. DOI: 10.1002/asi.20077 17

Bolter, J. (1991). Writing space: The computer, hypertext, and the history of writing. Hillsdale, NJ: Lawrence Erlbaum Associates. 21

Borgman, C. and Furner, J. (2002). Scholarly communication and bibliometrics. *Annual Review of Information Science and Technology*, 36, 3–72. DOI: 10.1002/aris.1440360102 17

boyd, d. (2007). *Why Youth (Heart) Social Network Sites: The Role of Networked Publics in Teenage Social Life*. MacArthur Foundation Series on Digital Learning - Youth, Identity, and Digital Media Volume (ed. David Buckingham). Cambridge, MA: MIT Press, pp. 119–142. 79

Brown, J. and Duguid, P. (2000). *The social life of information*. Boston: Harvard Business School Press. 20, 63

Brunk, B. (2001). Exoinformation and Interface Design. *ASIST Bulletin*, August / September 2001, 27(6), `http://www.asis.org/Bulletin/Aug-01/brunk.html`. 67

Buckland, M. (1991a). Information as thing. *Journal of the American Society of Information Science* 42:5. 351–360. DOI: 10.1002/(SICI)1097-4571(199106)42:5%3C351::AID-ASI5%3E3.0.CO;2-3 27

Buckland, M. (1991b). *Information and information systems*. New York: Praeger. 4

Buckland, M. (1997). What is a document? *Journal of the American Society of Information Science and Technology* 48, no. 9. 804–809.
DOI: 10.1002/(SICI)1097-4571(199709)48:9%3C804::AID-ASI5%3E3.0.CO;2-V 27

Bush, V. (1945). As We May Think. *The Atlantic Monthly*; July 1945, Volume 176, No. 1; 101–108. DOI: 10.1145/1113634.1113638 41

Christel, M. (2009). *Automated metadata in multimedia information systems: Creation, refinement, use in surrogates, and evaluation*. Synthesis Lectures on Information Concepts, Retrieval, and Services. ed. Gary Marchionini, no. 2. San Rafael: Morgan and Claypool Publishers. DOI: 10.2200/S00167ED1V01Y200812ICR002 37

Clark, A. (1997). Being there: Putting brain, body, and world together again. Cambridge, MA: MIT Press. 16

Cronin, B. (1995). *The scholar's courtesy: The role of acknowledgement in the primary communication process*. London: Taylor Graham. 17

Deek, F. and McHugh, J. (2008). *Open Source: Technology and policy*. NY: Cambridge University Press. 42

Dibbell, J. (1996). A rape in cyberspace: How an evil clown, a Haitian trickster spirit, two wizards, and a cast of dozens turned a database into a society. In M. Stefik (Ed.). *Internet dreams: Archetypes, myths, and metaphors*. Cambridge, MA: MIT Press. 58

Ding, W. and Lin, X. (2009). *Information Architecture: The Design and Integration of Information Spaces*. Synthesis Lectures on Information Concepts, Retrieval, and Services. ed. Gary Marchionini, no. 9. San Rafael: Morgan and Claypool Publishers.
DOI: 10.2200/S00214ED1V01Y200910ICR008 18

Dominick, J. (2004). *The IN-SITU Study of an Electronic Textbook in an Educational Setting*. Unpublished doctoral dissertation. Chapel Hill, NC: The University of North Carolina at Chapel Hill, 2004. 26

Eisenberg, B. M. (2008). Information Literacy: Essential Skills for the Information Age. *Journal of Library and Information Technology, Vol. 28,* No. 2, March 2008, pp. 39–47 21

Engelbart, D. (1963). A conceptual framework for augmentation of man's intellect. In P. Howerton and D. Weeks, *Vistas in information handling* (Vol I). Washington, DC: Spartan Books, p. 1–29. 13, 41

Frankel, P., Miller, F., and Paul, J. (2005). *Personal identity*. Cambridge, UK: Cambridge U. Press. 63

Friston, K. (2009). The free-energy principle: A rough guide to the brain? *Trends in Cognitive Sciences*. 13(7), 293–301. DOI: 10.1016/j.tics.2009.04.005 16

Furner. J. (2004). Information studies without information. *Library Trends*, 52(3), 427–446. 5

Gemmell, G., Bell, G., and Lueder, R. (2006). MyLifeBits: a personal database for everything. *Communications of the ACM*, vol. 49, no. 1, pp. 88–95. DOI: 10.1145/1107458.1107460 33

Glimcher, P., Camerer, C., Fehr, E., and Poldrack, R. (2009). Neuroeconomics: Decision making and the brain. London: Academic Press. 73

Hall, S. (2010). *Wisdom: From Philosophy to Neuroscience*. NY: Alfred Knopf. 8, 16

Hartzog, W. (2009). The privacy box: A software proposal. *First Monday*, 14(11). `http://firstmonday.org/htbin/cgiwrap/bin/ojs/index.php/fm/article/view/2682/2361`. 74

Hey, T., Tansley, S., and Tolle, K. (2009). *The fourth paradigm: Data-intensive scientific discovery*. Redmond, WA: Microsoft Research. `http://research.microsoft.com/en-us/collaboration/fourthparadigm/4th_paradigm_book_complete_lr.pdf`. 42

Hjorland, B. (1995). Toward a new horizon in information science: Domain-Analysis. Journal of the American Society for Information Science. 46(6), 400–425.
DOI: 10.1002/(SICI)1097-4571(199507)46:6%3C400::AID-ASI2%3E3.0.CO;2-Y 5

Hobart, M. and Schiffman, Z. (1998). *Information ages: Literacy, numeracy, and the computer revolution*. Baltimore, MD: Johns Hopkins University Press. 5

Hodge, G. (2000). Best practices for digital archiving: An information life cycle approach. *D-Lib Magazine*, January, 2000 (Vol 6, No. 1). http://dlib.org/dlib/january00/01hodge.html. DOI: 10.3998/3336451.0005.406 31

Hoffman, R., Sobel, M., and Teute, F. (1997). *Through a glass darkly: Reflections on personal identity in early America*. Chapel Hill, NC: UNC Press. 64

Homans, G. (1958). Social behavior as exchange. *American Journal of Sociology* 63(6): 597–606. DOI: 10.1086/222355 20

Hughes, J., Loww, S., and Sabat, S. (2006). Dementia: Mind, meaning, and the person. Oxford, UK: Oxford U. Press. pp. 1–39. 64

Hutchins, E. (1995) *Cognition in the Wild*. Cambridge, MA: MIT Press). 15, 49

Jansen, B. (2009). *Understanding User-Web Interactions via Web Analytics*. Synthesis Lectures on Information Concepts, Retrieval, and Services. ed. Gary Marchionini, no. 7. San Rafael: Morgan and Claypool Publishers. 60

Johnson, M. (1987). *The body in the mind*. Chicago : University of Chicago Press. 16

Johnson-Laird, P. (1983). *Mental models: Toward a cognitive science of language, inference and consciousness*. Harvard University Press. 11

Jones, W. *Future directions in personal information management*. Synthesis Lectures on Information Concepts, Retrieval, and Services. ed. Gary Marchionini, San Rafael: Morgan and Claypool Publishers. 34

Jones, W. and Teevan, J. (Eds.) (2007). *Personal information management:* Seattle, WA: University of Washington Press. 34

Kamm, F. (2005). Moral status and personal identity: Clones, embryos, and future generations. In Frankel, P., Miller, F., and Paul, J. (Eds.). *Personal identity*. Cambridge, UK: Cambridge U. Press. pp. 308–373. 64

Karamuftuoglu, M. (2009). Situating logic and information in information science. *Journal of the American Society for Information Science and Technology* 60(10), 2019–2031. DOI: 10.1002/asi.21108 5

Kelly, M. (2009). Memories of Friends Departed Endure on Facebook. *The Facebook Blog*. http://blog.facebook.com/blog.php?post=163091042130. (October 26, 2009). 40

Knudsen, E. (2007). Fundamental components of attention. *Annual review of neuroscience* 30: 57–78. DOI: 10.1146/annurev.neuro.30.051606.094256

Koenig, M. and McInerney, C. (in press). *Knowledge management processes in organizations: Theoretical Foundations and examples of practice.* Synthesis Lectures on Information Concepts, Retrieval, and Services. ed. Gary Marchionini, San Rafael: Morgan and Claypool Publishers. 8

Lalmas, M. (2009). *XML retrieval.* Synthesis Lectures on Information Concepts, Retrieval, and Services. ed. Gary Marchionini, no. 6, San Rafael: Morgan and Claypool Publishers. DOI: 10.2200/S00203ED1V01Y200907ICR007

Lanier, J. (2010). *You are not a gadget: A manifesto.* NY: Alfred Knopf. 31, 73, 78

Lankes, D. (2009). *New concepts in digital reference.* Synthesis Lectures on Information Concepts, Retrieval, and Services. ed. Gary Marchionini, no. 1. San Rafael: Morgan and Claypool Publishers. DOI: 10.2200/S00166ED1V01Y200812ICR001 37

Laughlin, S., de Ruyter van Steveninck, R., and Anderson, J. (1998). The metabolic cost of neural information. *Nature Neuroscience,* 1(1), 36–14. DOI: 10.1038/236 48

Leavitt, N. (2010). Network-usage changes push Internet traffic to the edge. *Computer,* 43(10) October, 13–15. DOI: 10.1109/MC.2010.293 27

Lessig, L. (2008). *Remix: Making art and commerce thrive in the hybrid economy.* NY: The Penguin Press. 31, 78

Lombard, M. and Ditton, T. (1997). At the heart of it all: The concept of presence. *Journal of Computer-Mediated Communication,* 3(2). `http://jcmc.indiana.edu/vol3/issue2/lombard.html`. DOI: 10.1111/j.1083-6101.1997.tb00072.x 57

Lose, R. (2010). *Information from processes: The science of information.* `http://InformationFromProcesses.org`. 4

Lose, R. (1997). A discipline independent definition of information. *Journal of the American Society for Information Science,* 48(3), 254–269.
DOI: 10.1002/(SICI)1097-4571(199703)48:3%3C254::AID-ASI6%3E3.0.CO;2-W 4

Machlup, F. (1962). *The production and distribution of knowledge in the United States.* Princeton, NJ: Princeton U. Press. 1

Marchionini, G. (2008). Human Information Interaction Research and Development. *Library and Information Science Research* 30(3), 165–174. DOI: 10.1016/j.lisr.2008.07.001 55

Marchionini, G., Shah, C., Lee, C., and Capra, R. (2009). Query Parameters for Harvesting Digital Video and Associated Contextual Information. *Proceedings of the ACM/IEEE Joint Conference on Digital Libraries.* (Austin, TX, June 15–19, 2009). pp. 77–86. DOI: 10.1145/1555400.1555414 40

Marshall, C. (2009). *Reading and writing the electronic book.*. Synthesis Lectures on Information Concepts, Retrieval, and Services. ed. Gary Marchionini, no. 9. San Rafael: Morgan and Claypool Publishers. DOI: 10.2200/S00215ED1V01Y200907ICR009 34

Marshall, C. and Bly, S. (2004). Sharing encountered information: digital libraries get a social life. In *Proceedings of the 4th ACM/IEEE-CS Joint Conference on Digital Libraries(JCDL '04).* Tucson, AZ. 218–227. NY: ACM Press. DOI: 10.1109/JCDL.2004.240018 20

Maslow, A. (1943). A Theory of Human Motivation. *Psychological Review* 50(4), 370–96. DOI: 10.1037/h0054346 18

Mayer-Schönberger, V. (2009). *Delete: The virtue of forgetting in the digital age.* Princeton, NJ: Princeton U. Press. 13

McLuhan, M. (1994). *Understanding media: the extensions of man.* Cambridge, MA: MIT Press.

Meho, L. and Sugimoto, C. (2009). Assessing the scholarly impact of information studies: A tale of two citation databases—Scopus and Web of Science. *Journal of the American Society for Information Science and Technology.* 60(12), 2499–2508. DOI: 10.1002/asi.21165 72

Miller, G. (1956). The Magical Number Seven, Plus or Minus Two. *The Psychological Review*, 1956, vol. 63, Issue 2, pp. 81–97. DOI: 10.1037/h0043158

Miller, P. (2004). *Rhythm Science.* (Mediaworks Pamphlets). The MIT Press. 33

Mitchell, T., Shinkareva, A., Carlson, A., Chang, K., Malave, V., Mason, R., and Just, M. (2008). Predicting human brain activity associated with the meaning of nouns. *Science*, 320, May 20, 2008. 1191–1195. DOI: 10.1126/science.1152876 45

Mitchell, W.J. (2003). *Me++ The cyborg self and the networked city.* Cambridge, MA: MIT Press. 56

Moores, T. (2005). Do consumers understand the role of privacy seals in e-commerce? *Communications of the ACM* 48(3), 86–91. DOI: 10.1145/1047671.1047674 75

Morris, M. and Teevan, J. (2009). *Collaborative Web Search: Who, What, Where, When, and Why.* Synthesis Lectures on Information Concepts, Retrieval, and Services. ed. Gary Marchionini, no. 12. San Rafael: Morgan and Claypool Publishers. DOI: 10.2200/S00230ED1V01Y200912ICR014 42

Muhlhausler, P. and Harre, R. (1990). *Pronouns and people: The linguistic construction of social and personal identity.* Okford, UK: Basil Blackwell. 64

Newell, A., and Simon, H. A. (1972). *Human problem solving.* Englewood Cliffs, NJ: Prentice-Hall. 13

Paas, F., Renkl, A., and Sweller, J. (2003). Cognitive load theory and instructional design: Recent developments. *Educational Psychologist, 38*, 1–4. DOI: 10.1207/S15326985EP3801_1 48

Pentland,A. (2008). *Honest signals: How they shape our world.* Cambridge, MA: MIT Press. 33

Posner, M. (1989). *Foundations of Cognitive Science.* Cambridge, MA: MIT Press. 301–356.

Quillian, R. (1967), Word Concepts: A theory and simulation of some basic semantic capabilities, *Behavioral Science*, 12: 410–430. DOI: 10.1002/bs.3830120511 15

Raichle, M. and Mintun, M. (2006). Brain work and brain imaging.*Annual Review of Neuroscience*, 29, 449–476. DOI: 10.1146/annurev.neuro.29.051605.112819 14, 15

Rajasekar, A., Moore, R., Hou, C., Lee, C., Marciano, R., de Torcy, A., Wan, M., Schroeder, W., Chen, S., Gilbert, L., Tooby, P., and Zhu, B. (2010). *iRODS Primer: Integrated Rule-Oriented Data System.* Synthesis Lectures on Information Concepts, Retrieval, and Services. ed. Gary Marchionini, no. 14. San Rafael: Morgan and Claypool Publishers. DOI: 10.2200/S00233ED1V01Y200912ICR012 38

Raymond, E. (1999). *The cathedral and the bazaar: Musings on Linux and open source by an accidental revolutionary.* Sebastopol, CA: O'Reilly and Associates Inc. 42

Rayner, K. (1998). Eye movements in reading and information processing: 20 years of research. *Psychological Bulletin*, 124: 372–422. DOI: 10.1037/0033-2909.124.3.372 15

Reeves, B. and Nass, C. (1996). *The media equation: How people treat computers, television, and new media like real people and places.* New York: Cambridge University Press. 23, 63

Renear, A. and Palmer, C. (2009). Strategic Reading, Ontologies, and the Future of Scientific Publishing.*Science.* Vol. 325, no. 5942, pp. 828–832. DOI: 10.1126/science.1157784 34

Resnikoff, H. (1989). *The illusion of reality.* New York: Springer-Verlag. 4

Roederer, J. (2005). *Information and its Role in Nature*, Heidelberg: Springer-Verlag. 47

Rosenfeld, L. and Morville, P. (2002). *Information architecture for the World Wide Web* (2nd Ed.).Sepastopol, CA: O'Reilly. 18

Ruger, S. (2009). *Multimedia information retrieval.* Synthesis Lectures on Information Concepts, Retrieval, and Services. ed. Gary Marchionini, no. 11. San Rafael: Morgan and Claypool Publishers. DOI: 10.2200/S00244ED1V01Y200912ICR010 37

Sawahata, Y. and Aizawa, K. (2003). Indexing of personal video captured by a wearable imaging system. (In E. Bakker, T. Huang, M. Lew, N. Sebe, and X. Zhou, Eds.) *Image and video retrieval: 2nd International Conference, CIVR 2003. Lecture Notes in Computer Science 2728.* Berlin: Springer. 342–351. 33

Schrader, A. (1984). In search of a name: Information science and its conceptual antecedents. *Library and Information Science Research*, vol 6., 227–271. 4

Sejnowski, T. and Smith Churchland, P. (1989). Brain and cognition. In M. Posner (Ed.), *Foundations of Cognitive Science.* Cambridge, MA: MIT Press. 301–356. 15

Sellen, A. and Whittaker, S. (2010). Beyond total capture: A constructive critique of lifelogging. *Communications of the ACM*, 53(5), 70–77. DOI: 10.1145/1735223.1735243 33

Shannon, C. (1948). A mathematical theory of communication, *Bell System Technical Journal*, Vol. 27, p. 379–423, 623–656. DOI: 10.1145/584091.584093 45, 47

Shannon, C. and Weaver, W. (1949). *The mathematical theory of communication.* The University of Illinois Press, Urbana, Illinois, 47

Shapiro, M. (2005). The identity of identity: Moral and legal aspects of technological self-transformation. In Frankel, P., Miller, F., and Paul, J. (Eds.). *Personal identity.* Cambridge, UK: Cambridge U. Press. pp. 308–373. 64

Shirky, C. (2010). *Cognitive surplus: Creativity and generosity in a connected age.* NY: Penguin Press. 78

Shneiderman, B. (2008). Science 2.0, *Science* 319(5868):1349–50. DOI: 10.1126/science.1153539 42

Standage, T. (1998). *The Victorian internet: The remarkable story of the telegraph and the nineteenth century's on-line pioneers.* New York: Walker and Company. 21

Sturm, B. (2000). The story listening trance experience,*Journal of American Folklore,* 113, 287–304. DOI: 10.2307/542104 52

Stutzman, F. and Kramer-Duffield, J. (2010). Friends Only: Examining a Privacy-Enhancing Behavior in Facebook. In *Proceedings of ACM CHI 2010.* Atlanta, GA, April 10–15, 2010. NY: ACM Press. DOI: 10.1145/1753326.1753559 20, 55

Sweller, J. (1988). Cognitive load during problem solving: Effects on learning. *Cognitive Science 12* (2), 257–285. DOI: 10.1207/s15516709cog1202_4 48

Thelwall, M. (2009). *Introduction to Webometrics: Quantitative Web Research for the Social Sciences.* Synthesis Lectures on Information Concepts, Retrieval, and Services. ed. Gary Marchionini, no. 4. San Rafael: Morgan and Claypool Publishers. DOI: 10.2200/S00176ED1V01Y200903ICR004 17

Thibaut, J., and Kelley, H. (1959). *The social psychology of groups.* New York: Wiley. 20

Tunkelang, D. (2009). *Faceted search*. Synthesis Lectures on Information Concepts, Retrieval, and Services. ed. Gary Marchionini, no. 5. San Rafael: Morgan and Claypool Publishers. DOI: 10.2200/S00190ED1V01Y200904ICR005

Turkle, S. (1995). *Life on the screen: Identity in the age of the Internet.* New York: Simon and Schuster. 58

Wang, F., Zeng, D., Hendler, J., Zhang, Q., Feng, Z., Gao, Y., Wang, H., and Lai, G. (2010). A study of the human flesh search engine: Crowd-powered expansion of online knowledge. *Computer*, 43(8), 45–53. DOI: 10.1109/MC.2010.216 79

White, R. and Roth, R. (2009).*Exploratory search: Beyond the query-response paradigm*. Synthesis Lectures on Information Concepts, Retrieval, and Services. ed. Gary Marchionini, no. 3. San Rafael: Morgan and Claypool Publishers. DOI: 10.2200/S00174ED1V01Y200901ICR003 38

Whitworth, B. and Whitworth, E. (2004). Spam and the social-technical gap.*Computer*, October, 2004. 38–45. DOI: 10.1109/MC.2004.177 19

Author's Biography

GARY MARCHIONINI

Gary Marchionini is the Cary C. Boshamer Professor in the School of Information and Library Science at the University of North Carolina at Chapel Hill. He received his PhD in mathematics education at Wayne State University. He is the author of more than 200 scholarly works and has had grants and contracts from the National Science Foundation, the Library of Congress, the National Cancer Institute, Microsoft, Google, and IBM among others. He currently serves as the dean of the School of Information and Library Science at UNC-CH.

Lightning Source UK Ltd.
Milton Keynes UK
10 March 2011

169038UK00001B/53/P